Springer Theses

Recognizing Outstanding Ph.D. Research

Aims and Scope

The series "Springer Theses" brings together a selection of the very best Ph.D. theses from around the world and across the physical sciences. Nominated and endorsed by two recognized specialists, each published volume has been selected for its scientific excellence and the high impact of its contents for the pertinent field of research. For greater accessibility to non-specialists, the published versions include an extended introduction, as well as a foreword by the student's supervisor explaining the special relevance of the work for the field. As a whole, the series will provide a valuable resource both for newcomers to the research fields described, and for other scientists seeking detailed background information on special questions. Finally, it provides an accredited documentation of the valuable contributions made by today's younger generation of scientists.

Theses may be nominated for publication in this series by heads of department at internationally leading universities or institutes and should fulfill all of the following criteria

- They must be written in good English.
- The topic should fall within the confines of Chemistry, Physics, Earth Sciences, Engineering and related interdisciplinary fields such as Materials, Nanoscience, Chemical Engineering, Complex Systems and Biophysics.
- The work reported in the thesis must represent a significant scientific advance.
- If the thesis includes previously published material, permission to reproduce this must be gained from the respective copyright holder (a maximum 30% of the thesis should be a verbatim reproduction from the author's previous publications).
- They must have been examined and passed during the 12 months prior to nomination.
- Each thesis should include a foreword by the supervisor outlining the significance of its content.
- The theses should have a clearly defined structure including an introduction accessible to new PhD students and scientists not expert in the relevant field.

Indexed by zbMATH.

Jason R. Thieman

Top Quark Spin Correlations with the CMS Detector

Doctoral Thesis accepted by Purdue University, USA

Springer

Jason R. Thieman
Northrop Grumman
Linthicum, MD, USA

ISSN 2190-5053 ISSN 2190-5061 (electronic)
Springer Theses
ISBN 978-3-031-76116-4 ISBN 978-3-031-76117-1 (eBook)
https://doi.org/10.1007/978-3-031-76117-1

This Springer imprint is published by the registered company Springer Nature Switzerland AG
The registered company address is: Gewerbestrasse 11, 6330 Cham, Switzerland

Supervisor's Foreword

The top quark, the heaviest known fundamental particle, provides a unique window into quantum chromodynamics (QCD) and other aspects of particle physics. Its short lifetime, which prevents hadronization, allows for the direct observation of its intrinsic properties, such as the top quark's polarization and spin correlations of top quark and top anti-quarks ($t\bar{t}$). These characteristics offer an unparalleled opportunity to test the Standard Model (SM) and probe for new physics. Top quark measurements open a window into better understanding the electroweak symmetry breaking due to the top quark's large Yukawa coupling to the Higgs boson field. The top quark is the only quark with a "natural" coupling to the Higgs boson of about 1. Hence, the mass of the top quark is most sensitive to the stability of the electroweak vacuum, and recent measurements indicate that it is meta-stable. Indications of new physics contributions and deviations from the SM in top quark production hence allow a better understanding of a variety of shortcomings in the current formulation of the SM.

In this thesis, Jason Thieman presents a comprehensive analysis of top quark polarization and top quark pair spin correlations using the full Run 2 data from the Compact Muon Solenoid (CMS) experiment at the Large Hadron Collider (LHC). The data, recorded at a center-of-mass energy of 13 TeV, include events containing oppositely charged leptons. For the first time at this energy, normalized differential cross sections sensitive to the complete polarization and spin-dependent parts of the $t\bar{t}$ production density matrix are measured, providing new insights into the properties of the top quark.

This work goes beyond a measurement of differential cross section distributions—it employs these results to search for new physics in an effective field theory (EFT) approach and also tests theoretical predictions, comparing them with simulations at next-to-leading-order (NLO) accuracy in QCD, along with electroweak corrections. All the findings are consistent with Standard Model predictions, and the measured distributions are used in the EFT approach to constrain anomalous chromomagnetic and chromoelectric dipole moments, as well as other higher dimension operators implementing hypothetical physics beyond the SM contributions. These precision measurements help to sharpen our understanding of fundamental interactions, particularly in a regime where quantum entanglement and spin correlations hold the key to quantum tomography—a new emerging field in top quark physics. The future of top quark studies at the LHC is incredibly promising. As we delve deeper into the large data sets made available by the High-Luminosity LHC (HL-LHC) upgrade, increasingly precise measurements of angular distributions will allow us to refine our models further and investigate subtle discrepancies in spin correlations.

Jason Thieman's work represents a significant contribution to these endeavors, advancing our understanding of top quark properties and their place in the larger framework of particle physics. It is a timely and valuable piece of research that not only reinforces the Standard Model but also sharpens the tools we use to look for physics beyond it. The careful and rigorous analysis within this thesis stands as a testament to the exciting times ahead in the field of high-energy physics.

West Lafayette, IN, USA Andreas Jung
September 2024

Acknowledgments

I express gratitude to my thesis advisor, Andreas Jung, for his guidance throughout my research. His expertise, encouragement, and feedback have been invaluable in shaping this work. I also thank Gabriele Benelli, Ron Lipton, and Corrinne Mills for the opportunities to collaborate under their supervision and learn from their expertise.

Acknowledgment is due to the other members of the top quark research group at Purdue, including Amandeep Bakshi, Giulia Negro, and Andrew Wildridge for their invaluable contributions to the measurements presented in this thesis. I would also like to recognize the top quark research group at DESY, Ajeeta Khatiwada, and Jacob Linacre for their instrumental role in developing the original measurement and unfolding framework that served as the foundation for this research.

My heartfelt appreciation is extended to the many colleagues and collaborators in the CMS experiment for their tireless efforts in maintaining and operating the CMS detector and for providing the high-quality data used for these measurements. Their exceptional expertise, hard work, and dedication have made this research possible.

I am also grateful to the members of my thesis committee, John Finley, Matthew Jones, and Martin Kruczenski, for their valuable time and constructive comments that have contributed to the quality and rigor of this dissertation.

Finally, I would like to thank my family and friends for their unwavering support, encouragement, and understanding throughout this challenging journey. Their love and care have been my source of strength and inspiration.

Introduction

Experiments to discover and determine the properties of Standard Model (SM) elementary particles are performed at particle accelerator facilities, such as the Large Hadron Collider (LHC) at the European Center for Nuclear Research (CERN), which is the largest and most powerful particle accelerator ever built. The LHC is capable of producing tens of millions of top quarks from high-energy proton-proton (pp) collisions every year, making it a "top quark factory." Due to its unique properties and role in exotic processes, the top quark is important for precision measurements of the SM and is sensitive to new physics (NP) beyond the SM (BSM). With a very short lifetime caused by its exceptionally large mass, the top quark decays before hadronizing, and its properties, including spin information, are transferred undiluted to its decay products and observable in their kinematic distributions.

This dissertation presents differential measurements of top quark (t and $\bar{t}$) polarizations and top quark pair ($t\bar{t}$) spin correlations, which probe the independent coefficients of the top-spin components of the $t\bar{t}$ production density matrix, using $137.7\text{fb}^{-1} \pm 1.6\%$ of data recorded by the CMS experiment at the LHC with $\sqrt{s} = 13\,\text{TeV}$, during 2016, 2017, and 2018. The measurements target all channels (e^+e^-, $e^\pm\mu^\mp$, $\mu^+\mu^-$) of the $t\bar{t}$ dileptonic decay mode with final states containing two oppositely charged leptons. The measurements follow the analysis strategy of CMS TOP-18-006 [1], published in *Physics Review D*, for which I collaborated as a co-author. Compared to CMS TOP-18-006, the measurements have been extended with additional observables, differential measurements as a function of $t\bar{t}$ invariant mass, several optimizations that reduce background contributions and systematic uncertainties, and significantly increased luminosity. With these optimizations and the inclusion of $t\bar{t}$ events the via τ decays as signal, signal purity has been increased from ~79% to ~93% and measurement precision for one-dimensional normalized unfolded cross-sections and extracted coefficients were improved by as much as a factor. All measured observables are corrected for detector efficiencies, acceptances, and migrations, unfolded to parton-level, and extrapolated to the full phase space using a regularized unfolding procedure with detector response obtained from MC simulated SM predictions with next-to-leading-order (NLO) matrix element accuracy interfaced with parton-shower and hadronization algorithms. The spin density coefficients are extracted from the unfolded distributions and compared to theoretical predictions, constituting a precision test of the SM. The measurement results can be used for a multitude of subsequent analyses:

- Effective field theory (EFT) interpretation to set limits on dimension-six operators relevant for hadronic $t\bar{t}$ production [1]
- Supersymmetry (SUSY) interpretation to set limits light supersymmetric top squark production [2]
- Observe entanglement in the $t\bar{t}$ system [3]
- Observe a violation of Bell's Inequality in the $t\bar{t}$ system [4]
- Search for evidence of Toponium near $t\bar{t}$ production threshold [5]

One of the ingredients for the $t\bar{t}$ spin correlation measurements is a set of scale factors correcting the trigger efficiencies in MC simulation to those observed in data. I performed detailed measurements of these scale factors, which were approved by the EGamma and Muon CMS Physics Object Groups and the CMS Top Quark Physics Analysis Group. The measurements were centrally provided and used by virtually all CMS Run II measurements analyzing $t\bar{t}$ events decaying via the dileptonic mode. In order to accommodate a search for Lorentz violation, I collaborated as a co-author and additionally provided dedicated measurements of the dilepton trigger efficiency scale factors as a function of sidereal time.

My contributions to the operation of the CMS experiment and to R&D efforts for the upcoming high luminosity LHC (HL-LHC) phase are many fold. I have spent many weeks in the CMS Remote Operations Center at Fermi National Accelerator Laboratory (FNAL) under the supervision of distinguished researcher Gabriele Benelli monitoring the data quality of the CMS silicon pixel and strip tracker sub-detectors and certifying collision data sets for tracking quality. During the long shutdown phase between Runs II and III, I was promoted to on-call expert, and later to shift leader, for the certification of ultra-legacy data sets. Moreover, I developed data quality monitoring tools for the monitoring of inefficient modules in the silicon strip sub-detector.

Also at FNAL, I contributed to the testing of HL-LHC pixel detector prototypes at the FNAL Test Beam Facility (FTBF) under the supervision of University of Illinois at Chicago physics professor Corrinne Mills. Besides data acquisition, I also contributed to the offline analysis of the data by identifying the reconstruction and alignment routines that were most effective and automating these routines in a workflow with built-in quality checks that allowed us to efficiently align the telescope and produce measurement results for large quantities of data. The results of the efficiency, resolution, charge response, and I-V curves of some of these prototypes were presented at the TREDI2020 Workshop on Advanced Silicon Radiation Detectors [6].

Collaborating under the guidance of FNAL R&D scientist Ron Lipton, we performed a series of TCAD and SPICE simulations to explore the plausibility of exploiting the initial transient induced current, described by the Ramo-Shockley Theorem, to achieve the timing precision required for four-dimensional tracking.

The results of these studies were presented at the ULITIMA fast timing conference at Argonne in October 2018 [7], and the conference proceedings were published to Nuclear Instrumentation and Methods in Physics [8].

The contents of this dissertation are outlined as follows:

- Chapter 1 contains a brief history of particle physics and a concise overview of the SM.
- Descriptions of the LHC and the CMS experiment are in Chap. 2.
- A summary of top quark physics at hadronic colliders, with particular emphasis on the spin properties of $t\bar{t}$ production and top quark decay, is set forth in Chap. 3.
- An overview of pp collision MC event simulation is given in Chap. 4.
- Chapter 5 contains details regarding object reconstruction and event selection, kinematic reconstruction of top quark pairs, background determination, event yields, and data sets used in the measurement.
- The unfolding procedure to obtain parton-level differential cross-sections extrapolated to the full phase space and sources of systematic uncertainties are explained in Chap. 6.
- Measurement results and the method for extracting spin density coefficients from the differential cross-sections are presented in Chap. 7.
- A summary and outlook are provided in Chap. 8.
- Appendix A contains a complete overview of trigger efficiency and corrective scale factor measurements that were used by virtually all CMS Run II measurements analyzing $t\bar{t}$ events decaying via the dileptonic mode.
- Simulation results investigating small pixels integrated with fast electronics for realizing the timing resolution required for four-dimensional tracking are presented in Appendix B.

References

1. CMS Collaboration, Measurement of the top quark polarization and $t\bar{t}$ spin correlations using dilepton final states in proton-proton collisions at $\sqrt{s} = 13$ TeV. Phys. Rev. D **100**(7), 072002 (2019). Preliminary results. https://doi.org/10.1103/PhysRevD.100.072002. arXiv: 1907.03729. [Online]. Available: https://cds.cern.ch/record/2681777
2. CMS Collaboration, Projection of the top quark spin correlation measurement and search for top squark pair production at the HL-LHC. CERN, Geneva, Tech. Rep. (2022). [Online]. Available: https://cds.cern.ch/record/2813262
3. Y. Afik, J.R.M. de Nova, Entanglement and quantum tomography with top quarks at the LHC. Eur. Phys. J. Plus **136**(9) (2021). [Online]. Available: https://doi.org/10.1140/epjp/s13360-021-01902-1
4. J.A. Aguilar-Saavedra, J.A. Casas, Improved tests of entanglement and bell inequalities with LHC tops. Eur. Phys. J. C **82**(8) (2022). [Online]. Available: https://doi.org/10.1140/epjc/s10052-022-10630-4
5. B. Fuks, K. Hagiwara, K. Ma, Y.-J. Zheng, Signatures of toponium formation in lhc run 2 data. Phys. Rev. D **104**, 034023 (2021). [Online]. Available: https://doi.org/10.1103/PhysRevD.104.034023

6. C. Mills, Performance of highly irradiated pixel sensors for the cms hl-lhc upgrade. TREDI2020: 15th "Trento" Workshop on Advanced Silicon Radiation Detectors, 2020. [Online]. Available: https://indico.cern.ch/event/813597/contributions/3727828/
7. J.T. Ronald Lipton, Fast timing with thick detectors and small pixels. ULITIMA 2018 (2018). [Online]. Available: https://indico.fnal.gov/event/24689/contributions/118309/attachments/76951/92176/TIming.pdf
8. R. Lipton, J. Theiman, Fast timing with induced current detectors. Nucl. Instrum. Methods Phys. Res. A: Accelerators Spectrometers Detectors Associat. Equipment **945**, 162423 (2019). ISSN: 0168-9002. https://doi.org/10.1016/j.nima.2019.162423. [Online]. Available: https://www.sciencedirect.com/science/article/pii/S0168900219309933

Contents

Chapter 1
Motivation and Theoretical Overview

Abstract This chapter sets the stage for understanding the subsequent experimental results by providing a brief historical context and a summary of the Standard Model (SM) of particle physics. Gaps in the SM are highlighted to motivate the ongoing search for new physics (NP) beyond the Standard Model (BSM).

1.1 Reductionism and a Brief History of Particle Physics

The reductionist approach to understanding the physical universe, by breaking it down to its smallest constituents and studying the interactions between them, is a central strategy of natural philosophy and the scientific method. Well before there was evidence of the discreteness of matter, and despite human senses suggesting its continuity, ancient Greek philosophers Democritus and Leucippus introduced the concept of the atom as the fundamental building block of matter in the fifth century B.C. [1]. The atomic theory of the ancient Greeks sought to reduce components of the cosmos to matter and empty space, propose that the fundamental constituents of matter were indivisible, and provide a natural explanation for the diversity of matter as resulting from atoms of different properties forming complex arrangements. This early reductionist philosophy was remarkably ahead of its time, and although ancient atomic theory and its concepts fell into obscurity after being rejected by later Greek philosophers, they were rediscovered by Enlightenment scientists in the early nineteenth century and laid the foundation for modern atomic theory, which forms the basis for understanding the structure of matter and the nature of chemical reactions.

Unprecedented progress in the field of particle physics has been made in the past two centuries following the adoption of the reductionist approach. After discovering periodic trends, Mendeleev published the first version of the periodic table in 1869 [2], which organized chemical elements in a tabular arrangement based on patterns of chemical properties and atomic weight. By the end of the nineteenth century, with the discoveries of the electron (e) by Thomson in 1897 [3] and nuclear radioactivity by Becquerel in 1896 [4], atoms were understood to not be indivisible and indestructible, and therefore could not be elementary constituents of matter. In

J. R. Thieman, *Top Quark Spin Correlations with the CMS Detector*,
Springer Theses, https://doi.org/10.1007/978-3-031-76117-1_1

the early twentieth century, Moseley revised the periodic table of elements after discovering atomic number [5], and this revision established a more accurate and consistent periodic table that led to a better understanding of the electron structure of atoms. Around the same time, Rutherford experimentally discovered the atomic nucleus [6] and formulated the concept of the proton (p). Some years later, to account for the missing mass discrepancy between measurements of atomic weight and atomic number, he also theorized the existence of neutrons. With the discovery of the neutron (n) in 1932 by Chadwick [7], for a time it was thought that protons, neutrons, and electrons constituted fundamental particles.

The development of a quantum mechanical framework in the twentieth century provided a deeper understanding of the behavior of atomic and subatomic particles that helped to illuminate previously unexplained phenomena such as atomic spectra and nuclear radioactivity. In the 1930s, Fermi formulated a theoretical framework that described beta decay as the result of the weak interaction between subatomic protons, neutrons, and electrons [8]. The development of the quantum field theory of electrodynamics (QED), led by Schwinger [9] and Feynman [10] in the 1940s and 1950s, provided a complete theoretical quantum framework for describing electromagnetism, which is responsible for phenomena such as electricity and light. With the development of cosmic ray experiments in the 1940s and particle accelerator experiments in the 1950s, an extensive "zoo" of subatomic and subnuclear particles were discovered. The formulation of the quark model by Gell-Mann in 1964 [11], and the development of quantum field theory of chromodynamics (QCD), elucidated that protons, neutrons, and many other species in the particle zoo did not constitute fundamental particles, but instead were composed of combinations of elementary quarks bound together by a strong nuclear force.

1.2 The Standard Model of Particle Physics

The Standard Model (SM) of particle physics (summarized in Fig. 1.1), developed through the 1960s and early 1970s, reduced the complex and diverse behavior of all atomic and subatomic particles to the elementary particles and the fundamental interactions between them. The SM is a quantum field theory (QFT), defined mathematically by the local gauge symmetry group $SU(3) \otimes SU(2) \otimes U(1)$, that incorporates the strong, weak, and electromagnetic interactions [12–18]. Field are entities described by Lagrangian densities ($\mathcal{L}$), defined over all space-time, that are able to produce quanta when excited. Elementary particles (described by their properties and quantum numbers spin, electric charge, color charge, and weak isospin), are understood to be quantized energy excitations of fields, truly indivisible point-like objects without any substructure whatsoever, that are both creatable and destructible.

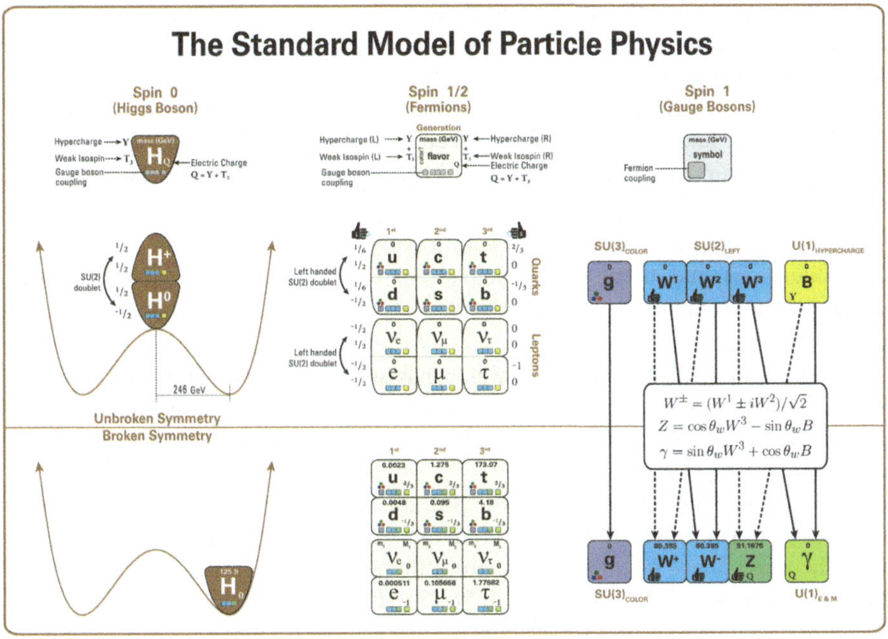

Fig. 1.1 The Standard Model of elementary particle physics [19]

1.2.1 *Fermions and Bosons*

In the SM, composite particles of matter are differentiated in terms of their constituent elementary particles of half-integer spin, called fermions, which are the building blocks of matter. Fermions are bound together by the exchange of fundamental, force-carrying particles of integer spin, called bosons.

Fermions are further categorized into the quark and lepton subgroups. Leptons and quarks carry electric charge (Q) and therefore participate in electromagnetic interactions. Quarks additionally carry red (R), green (G), or blue (B) color charge, and so can also participate in strong interactions. Because of this property and the nature of QCD, quarks cannot exist in isolation, but instead can only be observed confined to colorless bound states called hadrons.

There are three so-called "generations" of fermions, and each generation is composed of both a pair of leptons and a pair of quarks, resulting in a total of twelve fermions: six quarks and six leptons. Fermions of higher generations are more massive than those of the previous generation. The up (u) and down (d) quarks, together with the electron (e) and electron neutrino (ν_e) leptons, compose the first generation of fermions. Both protons and neutrons are composed of different combinations of three up and down quarks; therefore, virtually all atomic matter is composed of only first-generation fermions. The second generation of fermions consists of the charm (c) and strange (s) quarks as well as the muon (μ) and muon neutrino (ν_μ) leptons. Finally, the third generation of fermions is composed of the top (t) and bottom (b) quarks (sometimes nicknamed truth and beauty), together with the tau (τ) and tau neutrino (ν_τ) leptons.

The e, μ, and τ, referred to as charged leptons, carry electric charge $Q = -1$,[1] but the ν_e, ν_μ, and ν_τ neutrinos are electrically neutral. The u, c, and t quarks, referred to as "up-type," carry $Q = +\frac{2}{3}$, whereas the d, c, and b quarks, referred to as "down-type," carry $Q = -\frac{1}{3}$.

Left-handed quarks and leptons additionally carry weak isospin (I with third component I_3) and therefore participate in the weak interaction. Left-handed charged leptons and down-type quarks carry $I_3 = +\frac{1}{2}$, while left-handed neutrinos and up-type quarks $I_3 = -\frac{1}{2}$.

Every fermion, denoted arbitrarily as x, has an anti-matter counterpart denoted $\bar{x}$, with the same mass as x, but sign-inverted quantum numbers. The sign-inversions of red, green and blue color charges are anti-red (cyan), anti-green (magenta) and anti-blue (yellow), respectively.

The fundamental interactions of the SM are described by the exchange of force-mediating spin-1 vector bosons within the mathematical frameworks of gauge field theories. These mediators arise from the gauge field group generators of the symmetry groups $U(1)$, $SU(2)$, and $SU(3)$. The $U(1)$ group has one degree of freedom, and $SU(N)$ groups have $N^2 - 1$ degrees of freedom, so in total the SM has $1 + 3 + 8 = 12$ generators, corresponding to twelve gauge bosons. The gauge coupling constants of the fundamental interactions are free parameters that are determined experimentally. The SM also includes a spin-0 scalar boson associated with a scalar field responsible for the spontaneous symmetry breaking of the electroweak interaction and the generation of mass terms in the Lagrangian density for fermion fields.

1.2.2 The Electromagnetic Interaction

The photon (γ) is the massless and electrically neutral gauge boson that mediates the electromagnetic interaction between electrically charged particles. The electromagnetic interaction is described by QED with gauge symmetry group $U(1)$. The electromagnetic coupling, i.e. the fine structure constant ($\alpha_{EM} \simeq \frac{1}{137}$ at low energies), is significantly small relative to unity. For a given process, the momenta of mediating virtual photons depend on the energy scale, therefore the coupling constant "runs" and the interaction becomes stronger (but still significantly smaller than unity) at higher energy scales ($\alpha_{EM} \simeq \frac{1}{128}$ at scale $m_Z \simeq 91.2\,\text{GeV}$). As $\alpha_{EM} << 1$ even at high energy scales, perturbative power series expansions in α_{EM} have proven to be extremely effective at explaining experimental results with

[1] Electric charge is expressed in units of the elementary electric charge $e = 1.602176634 \times 10^{-19}\,\text{C}$.

QED. Despite only providing approximate predictions, this effectiveness has made QED one of the most well-tested and successful theories in physics.

1.2.3 The Weak Interaction

The $W^{\pm}$ and Z are massive gauge bosons that mediate the weak interaction between fermions carrying weak isospin (I with third component I_3). $W^{\pm}$ carry $Q = \pm 1$, while Z is electrically neutral. Weak isospins of $I_3 = \pm 1$ ($I_3 = 0$) are carried by the $W^{\pm}$ (Z) bosons. Their masses are $m_{W^{\pm}} \simeq 80.4\,\text{GeV}$ and $m_Z \simeq 91.2\,\text{GeV}$,[2] resulting in incredibly short lifetimes $\sim 10^{-25}$ s. The short lifetimes limit the interaction ranges to extremely small scales, because the distance these bosons travel before decaying is very short. The weak interaction is responsible for fermionic decay processes (such as nuclear radioactivity), is described in the SM by gauge symmetry group $SU(2)$, and is the only fundamental interaction that violates parity (P), charge conjugation parity (CP), and time-reversal (T) symmetries. Z bosons mediate the neutral weak interaction between same-flavor fermions, while $W^{\pm}$ bosons mediate the charged weak interaction between leptons of the same generation and can also flavor-mix quarks of different generations. Information about the conversion probabilities of quark flavor changing weak decays is encoded in the four free parameters (3 mixing angles + 1 CP violating phase) of the Cabibbo-Kobayashi-Maskawa (CKM) matrix [20], which cannot be determined from underlying principles and are measured experimentally.

1.2.4 The Strong Interaction

Gluons (g) are massless gauge bosons that mediate the strong interaction between color-charged particles. The strong interaction is described by QCD with gauge symmetry group $SU(3)_C$. While quarks (anti-quarks) carry color (anti-color) charge, gluons uniquely carry both color and anti-color charge, and therefore gluons can interact with other gluons at trilinear and quartic self-interaction vertices. There is a color octet of 8 different types of gluons, corresponding to 8 independent color state superpositions of the 3^2 color/anti-color combinations. One additional independent superposition exists mathematically, a color singlet carrying no color charge, but there is no experimental evidence for it and it is forbidden by $SU(3)_C$.

[2] In this dissertation, some quantities are expressed, when conventional, using the system of natural units. With this system, many expressions are simplified by setting the speed of light (c) and the reduced Planck's constant ($\hbar$) to unity and dimensional quantities can be expressed in units of energy or inverse energy. When these units are used, the standard dimensions can be recovered by appropriately multiplying or dividing by c and $\hbar$.

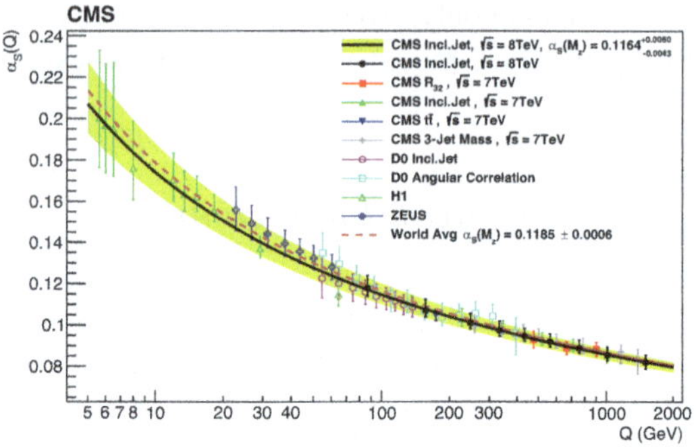

Fig. 1.2 The running of the strong coupling constant α_S as a function of the momentum transfer $Q = p_T$ [25]

The strong coupling α_S runs vigorously as the energy scale decreases, i.e. the interaction strength between quarks becomes stronger the further the quarks are separated from each other. This property of QCD results in color confinement [21], because before the energy between two separated quarks becomes arbitrarily large, a critical energy is reached for hadronization, in which the color-field lines snap and a new quark/anti-quark pair is created from the energy of the mediating gluon. Hadrons are classified by the flavor combinations of their valence quarks, but because quark interactions are mediated by gluons that can interact with themselves and split into quark/anti-quark pairs, there also exists a sea of virtual quarks and gluons within hadrons. Colorless flavor combinations of three valence quarks, such as protons and neutrons, are categorized as baryons, whereas colorless flavor combinations of one valence quark and one valence anti-quark, such as pions and kaons, are categorized as mesons. Experimental evidence for colorless flavor combinations of four valence quarks and one valence anti-quark, categorized as pentaquarks [22], and colorless flavor combinations of two valence quarks and two valence anti-quarks, categorized as tetraquarks [23] have also been reported in recent years. Of all known hadrons, only the proton appears to be stable; with a mean lifetime measured to be at least 10^{34} years, proton decay has never been observed.

Another consequence of the running of α_S is asymptotic freedom [18]. At higher and higher energy scales, α_S asymptotically approaches zero (see Fig. 1.2), i.e. quarks and gluons behave more like free particles at high energies. At sufficiently high energy scales, such as those in hard scattering processes at hadron collider experiments, perturbative power series expansions in α_S become effective at explaining experimental results with QCD. For example, $\alpha_S(Q = m_Z \simeq 91.2\,\text{GeV}) = 0.1171 \pm 0.0010$ [24].

1.2.5 The GWS Model, EWSB, and the BEH Mechanism

At low energy scales, the electromagnetic and weak interactions are distinctly separate forces. For energies beyond the electroweak transition scale $\sim$160 GeV, electroweak theory unifies these two interactions into a single force described by a spontaneously broken quantum field with gauge symmetry group $SU(2)_L \otimes U(1)_Y$, generated by weak isospin carried by left-handed fermion doublets (L) and weak hypercharge (Y). In the Glashow-Weinberg-Salam (GWS) model [12–14], the $U(1)_Y$ group is generated by weak hypercharge and mediated by a massless B^0 boson, and the $SU(2)_L$ group is generated by weak isospin and mediated by massless W^1, W^2, and W^3 bosons.

In the SM, the γ, Z, and $W^{\pm}$ bosons arise from the Brout-Englert-Higgs (BEH) mechanism [15–17] spontaneously inducing electroweak symmetry breaking (EWSB) in the GWS model. Electric charge $Q = I_3 + \frac{1}{2}Y$ arises as a specific linear combination of weak isospin and hypercharge. Electrically neutral γ and Z bosons arise from mixed states of the B^0 and W^3 bosons:

$$\begin{pmatrix} \gamma \\ Z^0 \end{pmatrix} = \begin{pmatrix} \cos\theta_W & \sin\theta_W \\ -\sin\theta_W & \cos\theta_W \end{pmatrix} \begin{pmatrix} B \\ W_3 \end{pmatrix} \tag{1.1}$$

where θ_W is the Weinberg weak mixing angle related to the gauge coupling strengths of $SU(2)$ and $U(1)$. So the Z boson couples to both left-handed and right-handed fermions via W^3 and B^0 respectively. Electrically charged $W^{\pm}$ bosons arise from linear combinations of the W^1 and W^2 bosons:

$$W^{\pm} = \frac{1}{\sqrt{2}}\left(W_1 \mp i W_2\right). \tag{1.2}$$

and $W^{\pm}$ bosons couple to only left-handed fermions.

In the GWS model, local gauge symmetry prohibits mass terms in the Lagrangian for gauge bosons (Goldstone's theorem), but the $W^{\pm}$ and Z bosons are experimentally known to have large masses. To both resolve this discrepancy and induce EWSB, the BEH mechanism adds a complex scalar $SU(2)$ doublet field ϕ to the SM that permeates all of space:

$$\phi = \begin{pmatrix} \phi^+ \\ \phi^0 \end{pmatrix} \tag{1.3}$$

with self-interacting potential energy:

$$V(\phi) = \mu^2 \phi^\dagger \phi + \lambda \left(\phi^\dagger \phi\right)^2 \quad \lambda > 0 \tag{1.4}$$

The BEH potential, colloquially referred to as a "Mexican-hat" potential for $\mu^2 < 0$, satisfies $SU(2) \otimes U(1)$. The energy at the center of the potential is higher than at a ring of equivalent vacuum states with vacuum expectation energy $\langle 0|\phi|0\rangle = \sqrt{\frac{-\mu^2}{2\lambda}} \equiv \frac{v}{\sqrt{2}}$ that breaks the electroweak gauge symmetry. By introducing the real and physical scalar Higgs field H, ϕ can be parameterized with an expansion around the vacuum expectation energy, and with a transformation to the unitary gauge:

$$\phi = \frac{1}{\sqrt{2}} \begin{pmatrix} \omega_2/2 + i\omega_1/2 \\ v + H + i\omega_3/2 \end{pmatrix} = \frac{e^{\frac{i\omega_j \cdot \sigma^j}{2v}}}{\sqrt{2}} \begin{pmatrix} 0 \\ v + H \end{pmatrix} \longrightarrow \phi' = \frac{1}{\sqrt{2}} \begin{pmatrix} 0 \\ v + H \end{pmatrix} \tag{1.5}$$

where σ_j are the Pauli matrices. Of the original four degrees of freedom in the complex scalar $SU(2)$ doublet field, the electrically charged ω_1 and ω_2 are absorbed to generate mass terms in the Lagrangian for the $W^{\pm}$ bosons $m_{W^{\pm}} = \frac{v}{2}$, while the electrically neutral ω_3 is absorbed to generate mass for the Z boson, such that $m_{W^{\pm}} = m_Z \cos\theta_W$. The remaining electrically neutral degree of freedom is split into v and the new field H, with quantum numbers $I_3 = -\frac{1}{2}$ and $Y = +1$ ($Q = 0$), which mediates interactions between the fermions and the BEH field via the scalar Higgs (H^0) boson, endowing them with mass $m_f = \frac{v}{\sqrt{2}} y_f$ proportional to their Yukawa coupling y_f and the vacuum expectation energy v. The nine[3] Yukawa couplings cannot be determined from any underlying principles and are free parameters that are determined experimentally. The Higgs boson also acquires mass $m_H = v\sqrt{2\lambda}$ through self-interaction with the BEH field. Both μ and λ are free parameters of the SM that determine $v \simeq 246$ GeV and m_H. The Higgs boson was experimentally discovered, in 2012 jointly by the CMS [27] and ATLAS [28] LHC experiments at CERN, with mass $m_H \simeq 125$ GeV. Peter Higgs and François Englert received the 2013 Nobel Prize in Physics for their theoretical predictions, while Robert Brout passed away in 2011.

1.2.6 Standard Model Lagrangian Density

In summary, the notation for the SM Lagrangian density can be abbreviated into the following form:

$$\mathcal{L}_{SM} = \mathcal{L}_{Yang-Mills} + \mathcal{L}_{Dirac} + \mathcal{L}_{Yukawa} + \mathcal{L}_{Higgs} \tag{1.6}$$

$$\mathcal{L}_{SM} = -\frac{1}{4} F_{\mu\nu} F^{\mu\nu} + i\bar{\psi}\not{D}\psi + \text{h.c.} + \bar{\psi}_i y_{ij} \psi_j \phi + \text{h.c.} + \left|D_\mu \phi\right|^2 - V(\phi) \tag{1.7}$$

where the first term (Yang-Mills) describes the dynamics of the gauge fields responsible for the fundamental interactions, the second and third terms (Dirac) describe how these gauge fields interact with the fermion fields (with h.c. corresponding to the hermitian conjugate of the preceding term), the fourth and fifth terms (Yukawa) describe how the fermion fields obtain mass via interaction with the BEH field, the

[3] The SM assumes the three neutrinos are massless, but it has been observed experimentally that neutrinos undergo flavor oscillations as they propagate through space, implying non-zero mass differences [26].

sixth and seventh terms (Higgs) describes the BEH potential and the interaction of the BEH field with the gauge fields.

1.3 Beyond the Standard Model

The SM is a central pillar of modern particle physics; many of its predictions have been confirmed by innumerable experiments to exquisite precision, and it is considered to be one of the most mature and successful theories in all of science. Despite its success, there are a number of known physical phenomena that are not explained by the SM, and it is considered to be an effective theory up to some scale $\Lambda \approx$ TeV. All theoretical and experimental investigations that are beyond the current understanding of the Standard Model of particle physics are collectively referred to as Beyond the Standard Model (BSM).

Following the example of electroweak unification, theoretical frameworks that aim to unify the strong and electroweak interactions into a single, unified force are referred to as Grand Unification Theories (GUT). While the scale of electroweak unification is $\sim 10^2$ GeV, the scale of grand unification is estimated to be $\sim 10^{14}$–10^{16} GeV. Given the extreme energy requirements, it is unsurprising that no experimental evidence for GUT predictions has been found yet at energy scales accessible to modern particle accelerator experiments.

The theoretical explanation for the experimental observation of neutrino oscillations predicts that the neutrino masses are non-zero. However, the SM predicts that neutrinos should be massless particles.

The discovery of the Higgs boson with mass $m_H \simeq 125$ GeV has created an inconsistency in the SM known as the "Hierarchy Problem." Assuming that the SM is valid up to some energy scale, virtual loop corrections to the Higgs boson mass generate quadratic divergences. Calculating a predicted Higgs boson mass compatible with the observed value requires unnatural fine-tuning, highlighting the likelihood of New Physics (NP) between the electroweak scale and the Planck scale, i.e. the natural scale of quantum gravity.

Of the natural forces considered to be fundamental, only the gravitational interaction between massive objects is not described by the SM framework in a quantum field theory formulation. A theory that reduces all four physical interactions into a single, unified theory is colloquially referred to as the "Theory of Everything," and it is the holy grail of particle physics. Einstein's general theory of relativity (GR) [29] is a classical theory of gravitation that describes the force of gravity as resulting from the curvature of spacetime caused by the presence of massive objects. All attempts to unify the SM and GR have failed, and the two are considered to be separate and incompatible theories. A major focus of current theoretical physics research is centered around the formulation of a QFT that describes the gravitational interaction, but a common feature of these theories is that they are non-renormalizable. Theories of quantum gravity predict the existence of a massless, spin-2 gauge boson dubbed the graviton. Experimentally, however, gravity is $\sim 10^{24}$

times weaker than the weak interaction, and the lack of experimental sensitivity makes it extremely difficult to verify the predictions of these theories.

There also are astrophysical and cosmological observed phenomena that are not explained by the SM. For example, there is no explanation for the observed asymmetry between matter and anti-matter in the visible universe. Based on the assumption that the universe started with equal quantities of matter and anti-matter, the conditions required for baryogenesis (i.e. the generation of a cosmological excess of baryons over anti-baryons) include baryon number violation, CP-violation, and charge conjugation symmetry (C) violation. Sources of CP-violation in the SM via the weak interaction are not sufficient to account for the extreme asymmetry and baryon number is not violated whatsoever. CP-violation via the strong interaction is mathematically allowed in the SM, but the relevant parameters have been inexplicably measured to be extremely small (strong CP problem). Astronomical observations of gravitational lensing effects and in the velocity of rotation curves of spiral galaxies suggest that 85% of all observed matter in the universe is dark matter, an unknown form of matter that has its namesake in that it does not emit, absorb or reflect electromagnetic radiation, and thus can only be indirectly inferred through its gravitational effects on visible matter. Although dedicated dark matter detection experiments exist that search for Weakly Interacting Massive Particles (WIMPs), conclusive direct observation of dark matter WIMPs has not yet occurred. Cosmological observations of galaxies also indicate that the expansion of space is accelerating. The cause of this expansion is unknown, but it is attributed to an enigmatic form of energy, dubbed dark energy, that is thought to be responsible for 73% of the total energy density of the universe.

Many BSM extensions have been made that predict new elementary particles that, depending on their properties and stability, could explain dark matter. Perhaps the most popular of these extensions is Supersymmetry (SUSY), which predicts the existence of superpartners for every fermion and boson in the SM by adding terms to the Lagrangian and make it symmetric with respect to matter and field-mediating particles. SUSY can resolve the hierarchy problem by providing superpartner terms that cancel the quadratic divergences to the Higgs boson mass from heavy virtual particles. Despite compelling arguments for why the predicted superpartners could be within reach of LHC experiments, the lack of experimental evidence for SUSY has led to a resurgence of interest in alternative BSM models to resolve these open questions.

SM Effective Field Theory (SMEFT) is a model-independent extension of the Standard Model which uses a systematic product expansion in terms of higher-dimensional operators to describe the effects of NP at higher energy scales [30]. Due to its agnostic construction, SMEFT allows for a broad range of possibilities for NP. The SM, composed of mass dimension-2 and mass dimension-4 operators, is assumed to be an effective field theory that describes physical interactions up to some energy scale Λ. The SMEFT Lagrangian is expanded as a power series in Λ, with the contributions beyond the SM in the expansion parameterized as dimensionless "Wilson" coefficients $\alpha_k^{(n)}$ of the mass dimension-n operators $\mathcal{O}_k^{(n)}$:

$$\mathcal{L}_{\text{SM+EFT}} = \mathcal{L}_{\text{SM}}^{(4)} + \sum_k \frac{\alpha_k^{(5)}}{\Lambda} \mathcal{O}_k^{(5)} + \sum_k \frac{\alpha_k^{(6)}}{\Lambda^2} \mathcal{O}_k^{(6)} + \mathcal{O}(\frac{1}{\Lambda^3}). \tag{1.8}$$

Beyond the SM operators are suppressed by powers of Λ according to their mass dimension, and so the expansion is usually truncated after the leading contributions. Requiring Lorentz and gauge invariance only allows one dimension-5 operator, which violates lepton number and may explain the generation and mixing of neutrino masses [31], but otherwise the leading contributions are of mass dimension-6 [32].

References

1. S. Berryman, Ancient atomism, in *The Stanford Encyclopedia of Philosophy*, ed. by E.N. Zalta, U. Nodelman, Winter 2022 (Metaphysics Research Lab, Stanford University, Stanford, 2022)
2. D. Mendeleev, On the relationship of the properties of the elements to their atomic weights. Z. Chem. **12**, 405–406 (1869)
3. J.J. Thomson, XI. Cathode rays. Lond. Edinb. Dublin Philos. Mag. J. Sci. **44**(269), 293–316 (1897)
4. H. Becquerel, Sur les radiations émises par phosphorescence. C. R. Acad. Sci. Paris **122**, 420–421 (1896)
5. R.G. Egdell, E. Bruton, Henry moseley, x-ray spectroscopy and the periodic table. Philos. Trans. R. Soc. A: Math. Phys. Eng. Sci. **378**(2180), 20190302 (2020). [Online] Available: https://doi.org/10.1098/rsta.2019.0302
6. E. Rutherford, Lxxix. The scattering of α and β particles by matter and the structure of the atom. Lond. Edinb. Dublin Philos. Mag. J. Sci. **21**(125), 669–688 (1911)
7. J. Chadwick, Possible existence of a neutron. Nature **129**(3252), 312 (1932)
8. F.L. Wilson, Fermi's theory of beta decay. Am. J. Phys. **36**(12), 1150–1160 (1968). https://doi.org/10.1119/1.1974382
9. J. Schwinger, Quantum electrodynamics. I. A covariant formulation. Phys. Rev. **74**, 1439–1461 (1948). [Online] Available: https://link.aps.org/doi/10.1103/PhysRev.74.1439
10. R.P. Feynman, Space-time approach to quantum electrodynamics. Phys. Rev. **76**, 769–789 (1949). [Online] Available: https://link.aps.org/doi/10.1103/PhysRev.76.769
11. M. Gell-Mann, A schematic model of baryons and mesons. Phys. Lett. **8**(3), 214–215 (1964). ISSN: 0031-9163. https://doi.org/10.1016/S0031-9163(64)92001-3. [Online] Available: https://www.sciencedirect.com/science/article/pii/S0031916364920013
12. S.L. Glashow, Partial-symmetries of weak interactions. Nucl. Phys. **22**(4), 579–588 (1961). ISSN: 0029-5582. https://doi.org/10.1016/0029-5582(61)90469-2. [Online] Available: https://www.sciencedirect.com/science/article/pii/0029558261904692
13. S. Weinberg, A model of leptons. Phys. Rev. Lett. **19**, 1264–1266 (1967). [Online] Available: https://link.aps.org/doi/10.1103/PhysRevLett.19.1264
14. A. Salam, Weak and electromagnetic interactions, in *Selected Papers of Abdus Salam*, pp. 244–254. eprint: https://www.worldscientific.com/doi/pdf/10.1142/9789812795915_0034. [Online] Available: https://www.worldscientific.com/doi/abs/10.1142/9789812795915_0034
15. P. Higgs, Broken symmetries, massless particles and gauge fields. Phys. Lett. **12**(2), 132–133 (1964). ISSN: 0031-9163. https://doi.org/10.1016/0031-9163(64)91136-9. [Online] Available: https://www.sciencedirect.com/science/article/pii/0031916364911369
16. P.W. Higgs, Broken symmetries and the masses of gauge bosons. Phys. Rev. Lett. **13**, 508–509. [Online] Available: https://link.aps.org/doi/10.1103/PhysRevLett.13.508

17. F. Englert, R. Brout, Broken symmetry and the mass of gauge vector mesons. Phys. Rev. Lett. **13**, 321–323 (1964). [Online] Available: https://link.aps.org/doi/10.1103/PhysRevLett.13.321
18. D.J. Gross, F. Wilczek, Ultraviolet behavior of non-abelian gauge theories. Phys. Rev. Lett. **30**, 1343–1346 (1973). [Online] Available: https://link.aps.org/doi/10.1103/PhysRevLett.30.1343
19. L. Boyle, Standard model of particle physics–most complete diagram. [Online] Available: https://commons.m.wikimedia.org/wiki/File:Standard_Model_Of_Particle_Physics--Most_Complete_Diagram.png
20. M. Kobayashi, T. Maskawa, CP-violation in the renormalizable theory of weak interaction. Prog. Theor. Phys. **49**(2), 652–657 (1973). ISSN: 0033-068X. eprint: https://academic.oup.com/ptp/article-pdf/49/2/652/5257692/49-2-652.pdf. [Online] Available: https://doi.org/10.1143/PTP.49.652
21. H. Fritzsch, M. Gell-Mann, H. Leutwyler, Advantages of the color octet gluon picture. Phys. Lett. B **47**(4), 365–368 (1973). ISSN: 0370-2693. https://doi.org/10.1016/0370-2693(73)90625-4. [Online] Available: https://www.sciencedirect.com/science/article/pii/0370269373906254
22. LHCb Collaboration, Observation of $J/\psi p$ resonances consistent with pentaquark states in $\Lambda_b^0 \to J/\psi K^- p$ decays. Phys. Rev. Lett. **115**, 072001 (2015). [Online] Available: https://link.aps.org/doi/10.1103/PhysRevLett.115.072001
23. Belle Collaboration, Study of $e^+e^- \to \pi^+\pi^- J/\psi$ and observation of a charged charmoniumlike state at belle. Phys. Rev. Lett. **110**, 252002 (2013). [Online] Available: https://link.aps.org/doi/10.1103/PhysRevLett.110.252002
24. D. Boito, M. Golterman, K. Maltman, S. Peris, M.V. Rodrigues, W. Schaaf, Strong coupling from an improved τ vector isovector spectral function. Phys. Rev. D **103**, 034028 (2021). [Online] Available: https://link.aps.org/doi/10.1103/PhysRevD.103.034028
25. C. Collaboration, Summaries of CMS cross section measurements. [Online] Available: https://twiki.cern.ch/twiki/bin/view/CMSPublic/PhysicsResultsCombined
26. Super-Kamiokande Collaboration, Evidence for oscillation of atmospheric neutrinos. Phys. Rev. Lett. **81**, 1562–1567 (1998). [Online] Available: https://link.aps.org/doi/10.1103/PhysRevLett.81.1562
27. CMS Collaboration, Observation of a new boson at a mass of 125 GeV with the CMS experiment at the LHC. Phys. Lett. B **716**(1), 30–61 (2012). ISSN: 0370-2693. https://doi.org/10.1016/j.physletb.2012.08.021. [Online] Available: https://www.sciencedirect.com/science/article/pii/S0370269312008581
28. ATLAS Collaboration, Observation of a new particle in the search for the standard model Higgs boson with the ATLAS detector at the LHC. Phys. Lett. B **716**(1), 1–29 (2012). ISSN: 0370-2693. https://doi.org/10.1016/j.physletb.2012.08.020. [Online] Available: https://www.sciencedirect.com/science/article/pii/S037026931200857X
29. A. Einstein, Die feldgleichungen der gravitation. Sitzungsberichte der Königlich PreuSischen Akademie der Wissenschaften (Berlin), 844–847 (1915)
30. W. Buchmüller, D. Wyler, Effective lagrangian analysis of new interactions and flavour conservation. Nucl. Phys. B **268**(3), 621–653 (1986). ISSN: 0550-3213. https://doi.org/10.1016/0550-3213(86)90262-2. [Online] Available: https://www.sciencedirect.com/science/article/pii/0550321386902622
31. S. Weinberg, Baryon- and lepton-nonconserving processes. Phys. Rev. Lett. **43**, 1566–1570 (1979). [Online] Available: https://doi.org/10.1103/PhysRevLett.43.1566
32. B. Grzadkowski, M. Iskrzyski, M. Misiak, J. Rosiek, Dimension-six terms in the standard model lagrangian. J. High Energy Phys. **2010**(10), 85 (2010). ISSN: 1029-8479. [Online] Available: https://doi.org/10.1007/JHEP10(2010)085

Chapter 2
The Compact Muon Experiment at the Large Hadron Collider

Abstract This chapter describes the Large Hadron Collider (LHC) accelerator complex and the Compact Muon Solenoid (CMS) experiment. The design and operational principles of the LHC are discussed, emphasizing its role in the acceleration and collision of proton beams. Brief details of the CMS detector's subsystems, including the silicon tracker, calorimeters, and muon systems, are given to explain their roles in particle detection and identification. The CMS experiment's capabilities in testing SM predictions and exploring NP are highlighted.

2.1 The Large Hadron Collider

Experiments to discover and determine the properties of SM particles are performed at particle accelerator facilities. The Large Hadron Collider (LHC) is a 27 km circular particle accelerator located at the European Center for Nuclear Research (CERN) designed to accelerate two counter-rotating proton beams to energies up to 6.8 TeV (99.999999% the speed of light) and collide them at designated interaction points (IP). The LHC is the largest and most powerful particle accelerator ever built. Its design, construction, and operation are towering engineering achievements of mankind, requiring an international collaboration of thousands of scientists, engineers, and technicians from more than 100 countries. The successful operation of the LHC has contributed to our understanding of the fundamental nature of matter by providing conditions to test the hypotheses of the SM and enabling many important discoveries in particle physics. The primary design criteria for the LHC were to provide conditions for elucidating the nature of EWSB, testing the hypothesis of the BEH mechanism, and discovering the Higgs boson.

2.1.1 Probing Small Scales with High Energy

The De Broglie relation states that the wavelength of a particle is inversely proportional to its momentum [1]. According to Heisenberg's uncertainty principle,

J. R. Thieman, *Top Quark Spin Correlations with the CMS Detector*,
Springer Theses, https://doi.org/10.1007/978-3-031-76117-1_2

Table 2.1 In the twentieth century, particle accelerator energy increased many orders of magnitude from ~100 keV in the 1930s to ~10 TeV in the 2010s, and current distance scales probed are down to $\sim 10^{-19}$ m

Decade	Energy	Distance probed
1930	~100 keV	$\sim 10^{-11}$ m
1950	~100 MeV	$\sim 10^{-14}$ m
1970	~100 GeV	$\sim 10^{-17}$ m
1990	~1 TeV	$\sim 10^{-18}$ m
2010	~10 TeV	$\sim 10^{-19}$ m

the position of a particle with a short wavelength cannot be precisely determined [2]. Thus, the motivation for building higher energy particle colliders is to probe smaller distance scales. In the twentieth century, particle accelerator energy increased many orders of magnitude from ~100 keV in the 1930s to ~10 TeV in the 2010s (see Table 2.1), and current distance scales probed are down to $\sim 10^{-19}$ m.

2.1.2 Accelerator Complex at CERN

Before reaching the LHC, protons must be accelerated through a series of smaller accelerators at CERN's accelerator complex. A schematic of the LHC at CERN's accelerator complex in Geneva at the border of Switzerland and France is shown in Fig. 2.1. Protons are produced from hydrogen gas by inducing electric discharge in a duoplasmatron. The 80 m Linear Accelerator 2 (Linac 2) accelerates protons to an energy of 160 MeV before injecting them into the 157 m circumference Proton Synchrotron Booster (PSB), which divides the beam into bunches consisting of more than 100 billion protons with 25 ns separation, and further increases the proton energy to 2 GeV. Next, the proton bunches are injected into the 628 m circumference Proton Synchrotron (PS), which accelerates the bunches to 26 GeV before injecting them into the 7 km circumference Super Proton Synchrotron (SPS), which increases the proton energy to 450 GeV before finally injecting them into the LHC, which accelerates them to their final (2016, 2017, and 2018) energy of 6.5 TeV.

Ultra-high vacuum conditions $\sim 10^{-11}$ mbar are maintained in the LHC beam pipes and 16 radiofrequency (RF) cavities with electric fields oscillating at a frequency resonant with the bunching frequency are used to accelerate the protons. Superconducting magnets cooled with superfluid helium to temperatures below 2 K and operating with magnetic fields exceeding 8 T [4] are used to shape and control the beam directions. 1232 dipole magnets bend the beams around the ring and 474 quadrupole magnets are used to focus and collimate the beams. The 25 ns bunch separation corresponds to 2808 proton bunches simultaneously in the LHC. At the four LHC IPs, proton bunches cross at a rate of 40 MHz, creating collisions corresponding to temperatures a billion times hotter than the core of the sun. Final states after the collisions are measured and recorded by the CMS, ATLAS, ALICE, and LHCb particle detectors.

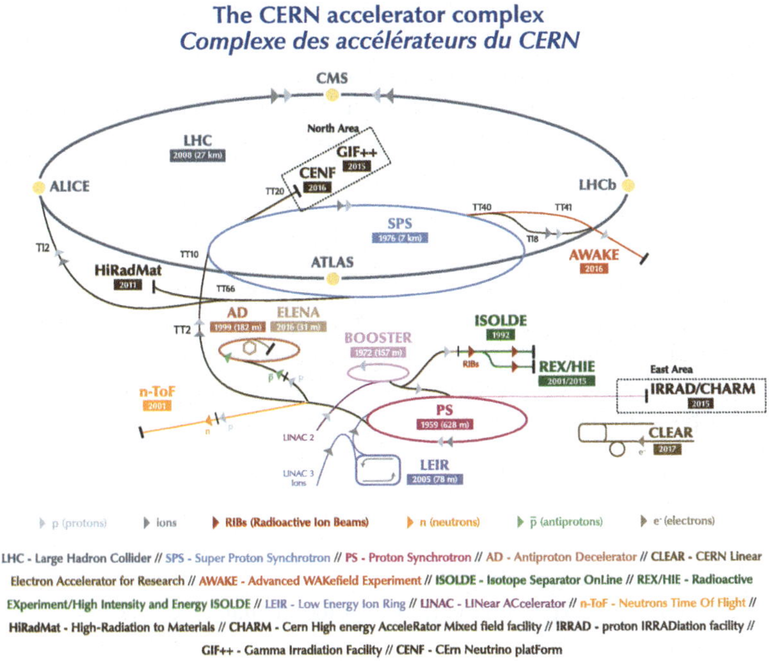

Fig. 2.1 A schematic of the LHC at CERN's accelerator complex in Geneva at the border of Switzerland and France [3]

2.1.3 *Luminosity*

When two particle beams cross paths in a particle accelerator, the number of events of a particular process that occur per second is given by:

$$\frac{dN_{\text{Events}}}{dt} = \mathcal{L}_{\text{Inst}} \cdot \sigma_{\text{Process}} \tag{2.1}$$

where $\mathcal{L}_{inst}$ is the instantaneous luminosity and σ_{Process} is the cross-section of the process under consideration. For the $\sqrt{s} = 13$ TeV LHC, the total cross-section for all inelastic processes has been measured to be $\sigma_{\text{Inelastic}} = 68.6 \pm 0.5$ (Syst) $\pm$ 1.6 (Lumi) mb [5]. The instantaneous luminosity can be parameterized [6] as:

$$\mathcal{L}_{\text{Inst}} = \frac{N_b^2 n_b f_{\text{rev}}}{A_{eff}} \tag{2.2}$$

where $N_b = 1.15 \times 10^{11}$ is the number of particles per bunch, $n_b = 2808$ the number of bunches per beam, $f_{\text{rev}} = 11{,}245$ Hz the revolution frequency in the collider, and $A_{eff} = \sqrt{\varepsilon_x \beta_x \varepsilon_y \beta_y}$ is the effective area of the luminous region, where

$\beta_{x,y}$ and $\varepsilon_{x,y}$ are the beam amplitude and emittance transverse to the beam axis, and is measured regularly during Van der Meer (VdM) scans. $\mathcal{L}_{\text{Inst}}$ decays over time due to the intensity degradation and beam emittance, with a mean life of $\sim$15 h [4]. The high $\mathcal{L}_{\text{Inst}}$ of the LHC typically results in many proton-proton interactions per bunch crossing. The number of collisions per bunch crossing is referred to as pile-up (PU), and is quantified as:

$$PU = \frac{\mathcal{L}_{\text{Inst}} \cdot \sigma_{\text{Inelastic}}}{n_b f_{\text{rev}}} \tag{2.3}$$

Typical PU in CMS recorded data from 2016 through 2018 was around 15–30 interaction vertices per bunch crossing, but high (low) extremes routinely occurred during the beginning (end) of beam fills due to instantaneous luminosity decay. For a particular process, the total number of events that occur in a recorded data set is given by:

$$N_{\text{Events}} = \mathcal{L} \cdot \sigma_{\text{Process}} \tag{2.4}$$

where the integrated luminosity $\mathcal{L} = \int \mathcal{L}_{\text{Inst}}\, dt$ is the instantaneous luminosity integrated over the corresponding periods of time the data was recorded.

2.1.4 *Upcoming LHC High Luminosity Phase*

During 2016, 2017, and 2018, the LHC operated with beam energies of 6.5 TeV, corresponding to center-of-mass collision energies of 13 TeV. Recently, the LHC finished upgrading the beam energies to 6.8 TeV. After operating at $\sqrt{s} = 13.6$ TeV through 2025, the LHC plans to vastly increase the instantaneous luminosity, and substantial R&D effort is underway to prepare accelerator and detector upgrades for high luminosity (HL-LHC) environments.

2.2 The Compact Muon Solenoid Detector

Particle detectors measure electrical signals and collect scintillated light produced by particles interacting with the material of the detector. The Compact Muon Solenoid (CMS) is a general-purpose detector, built for testing predictions of the SM, exploring models BSM, and discovering NP. The primary design benchmarks for CMS were precision studies of QCD, electroweak, and flavor physics, as well as the detection of the Higgs boson. CMS is a collection of sub-detectors that are layered cylindrically around the IP, with barrel and end-cap/disk components, that provide the detector with near-hermetic coverage of the solid angle. An exploded

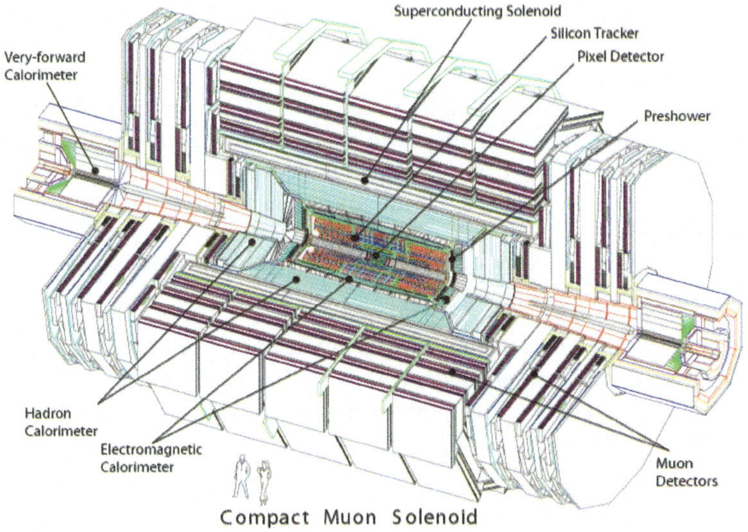

Fig. 2.2 An exploded view of the CMS detector layout with labeled sub-detectors [7]. The human figures are included for scale

view of the CMS detector layout with labeled sub-detectors is shown in Fig. 2.2. Overall, the CMS detector is 21.6 m long, with a diameter of 14.6 m, and weighs more than 12,500 t.

2.2.1 *Coordinate System*

The CMS origin is located at the nominal interaction point of the beams at the center of the experiment. The $\hat{y}$-axis points vertically upward, the $\hat{z}$-axis is along the beamline, and the $\hat{x}$-axis points towards the center of the LHC ring. When analyzing the kinematics of high-energy particle collisions, it is more conventional to use the $p_{\mathrm{T}}, \eta, \phi$ coordinate system:

- The transverse momentum $p_{\mathrm{T}} = \sqrt{{p_x}^2 + {p_y}^2}$ is the momentum component of a particle perpendicular to the beam axis.
- The pseudorapidity η is a measure of the angle between the particle's momentum vector and the beam axis. Pseudorapidity is defined as:

$$\eta = \frac{1}{2} \ln \frac{|\vec{p}| + p_z}{|\vec{p}| - p_z} = -\ln \left[\tan \left(\frac{\theta}{2} \right) \right] \tag{2.5}$$

in terms of the polar angle θ, which is measured from the z-axis. For particles with $m \lll |\vec{p}|$, it is sometimes preferable to use rapidity, defined as:

$$y = \frac{1}{2} \ln \frac{E + p_z}{E - p_z} \tag{2.6}$$

which converges to pseudorapidity in the ultra-relativistic limit. One advantage of using rapidity is that differences in y are Lorentz invariant for longitudinal Lorentz boosts along the beam axis. Differences in η are only Lorentz invariant for longitudinal Lorentz boosts in the ultra-relativistic limit, but it is easier to measure than y.
- The azimuthal angle ϕ is the angle in the transverse plane between the particle's momentum vector and the $\hat{x}$-axis.

The energy E of a particle, together with its momentum coordinates (p_T, η, ϕ), fully characterize the particle's momentum and energy in space and time.

A measure of angular separation between objects i and j can be defined using η and ϕ coordinates:

$$\Delta R_{ij} = \sqrt{(\phi_i - \phi_j)^2 + (\eta_i - \eta_j)^2}; \tag{2.7}$$

this quantity is useful for clustering particle showers into jets and rejecting events with overlapping objects.

2.2.2 *Solenoid Magnet*

Central to the design of the CMS detector, and the source of its name, is the 12.9 m-long, 5.9 m inner diameter [7], 220 t [8], niobium-titanium alloy superconducting solenoid, providing an axial and uniform 3.8 T magnetic field within the solenoid, and a return field outside the solenoid that is confined by an iron yoke that contains and guides the field. The solenoid is large enough that the inner tracking detectors and calorimeters fit inside, which advantageously eliminates energy losses caused by showering in the coil material before energy can be measured in the calorimeters. The magnetic field of the solenoid, depicted in Fig. 2.3, bends the paths of charged particles as they move through the detector, enabling the tracker to make good resolution measurements of their transverse momentum and charge from their curvature.

2.2.3 *Silicon Inner Tracker*

The inner tracker consists of silicon pixel detectors surrounded by silicon strip detectors. Silicon detector technology provides high granularity, fast response, radiation hardness, and efficient cooling, while also minimizing material to limit multiple scattering, bremsstrahlung, photon conversion, and nuclear interactions.

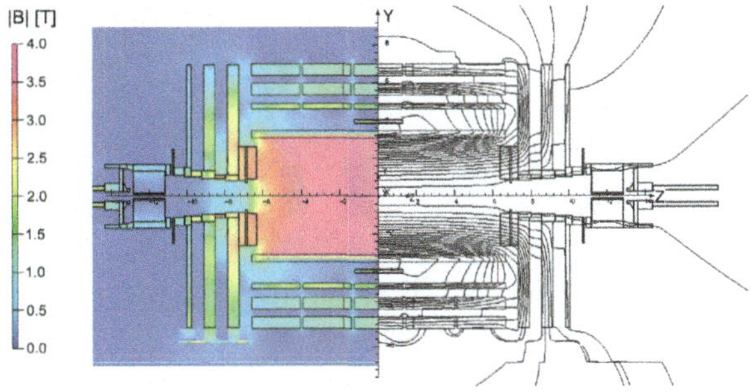

Fig. 2.3 Magnitude of magnetic field (Left) and field lines (Right) simulated on a longitudinal section of the CMS detector [9]

High-energy charged particles ionize the silicon as they traverse the detectors, leaving trails of position measurements that are used to reconstruct the particle trajectories with sophisticated tracking algorithms. The magnetic field of the solenoid exerts a force perpendicular to a charged particle's transverse momentum, and transverse momentum magnitude is measured from the curvature of its trajectory. The inner tracker cylindrical volume of length 5.8 m and diameter 2.6 m covers out to $|\eta| < 2.5$ and is separated into the barrel and forward regions [8].

The silicon pixel tracker operates closest to the IP, in the region with the highest radiation and particle flux from the *pp* collisions. The primary limiting factor for impact parameter resolution is the radial distance of the innermost tracking layer. In the barrel region, three layers of pixel detectors, 53 cm long, are at radial distances $r = 4.4, 7.3, 10.2$ cm from the beam, while in the forward regions, two end-cap disks with radii 6 and 15 cm are placed on each side at $z = \pm 34.5$ and 46.5 cm [8]. The doping scheme for the pixels is n-on-n, which can be operated partially depleted even at high fluence. To minimize occupancy in the inner region of high particle flux and achieve good vertex resolution for the identification of primary and secondary vertices, pixel detectors are highly granular and finely segmented in two dimensions with pixel size $\simeq 100 \times 150\ \mu\text{m}^2$ [8]. The pixel detector consists of 1440 modules that cover a total area of $\simeq$1 m^2, so altogether it contains about 66 million pixels [7]. In the barrel region, $r - \phi$ resolution is improved through charge-sharing among pixels due to the Lorentz effect, while in the forward region, modules are rotated in a turbine-like geometry to improve the resolution by inducing charge-sharing. The spatial resolution of the pixel detector is 10 μm (20 μm) for $r - \phi$ (z) measurements [8].

During the LHC winter shutdown between 2016 and 2017, the entire pixel tracker was replaced (see Fig. 2.4) with the Phase-1 upgrade [10]. A fourth layer was added in the barrel region, and the radius of the innermost layer was reduced from $r =$ 4.4 cm to $r = 2.9$ cm, a fifth layer was added in the forward regions, and the total number of pixels increased to 124 million. The cooling system was replaced with evaporative CO_2 cooling technology, which utilizes the latent heat of vaporization to efficiently remove heat. After the replacement, the silicon inner tracker altogether contributes total radiation length up to $\sim 0.5 X_0$ and total interaction length up to $\sim 1.8\lambda_I$, mostly from supports, cables, and cooling [11].

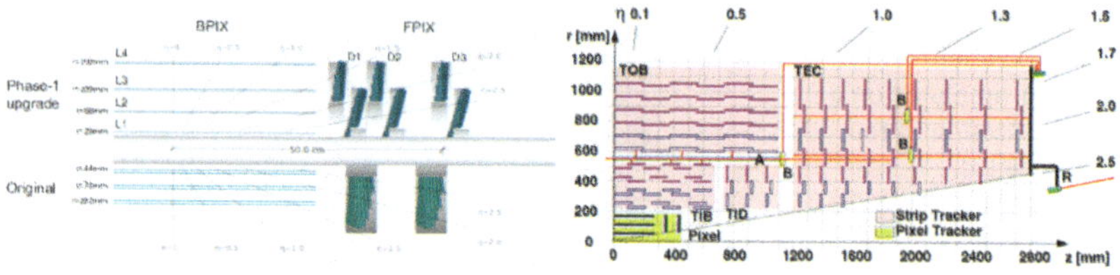

Fig. 2.4 Silicon Pixel Tracker (Left) before and after the Phase-1 upgrade [12]. Silicon Inner Tracker before Pixel Phase-1 upgrade (Right) [13]

Surrounding the pixel tracker is the silicon strip tracker (see Fig. 2.4), consisting of 15,148 modules with 24,244 silicon sensors, that cover a total area of $\simeq$198 m $^{-2}$, and altogether contains 9.6 million microstrips [8]. In the barrel region, the strip tracker is separated into the Tracker Inner Barrel (TIB), which covers from 20 cm $< r <$ 55 cm out to $z = \pm$65 cm, and the Tracker Outer Barrel (TOB), which covers from 55 cm $< r <$ 116 cm out to $z = \pm$118 cm [7]. In the forward regions, the strip tracker is separated into the Tracker Inner Disks (TID$^{\pm}$), which cover $\pm$65 cm $< z < \pm$124 cm and 20 cm $< r <$ 55 cm, and Tracker End-Caps (TEC$^{\pm}$), which cover $\pm$124 cm $< z < \pm$282 cm and 22.5 cm $< r <$ 113.5 cm [7]. The doping scheme for strips is p-on-n, which undergoes type inversion after sufficient radiation damage to the crystal lattice and change [8]. Strips are only finely segmented in one dimension, with minimum cell sizes of $\simeq$10 cm $\times$ 80 μm in the TIB and TID, and maximum cell sizes of $\simeq$25 cm $\times$ 180 μm in the TOB and TEC. The TIB (TOB) has four (six) layers of silicon microstrip detectors, providing up to four (six) $r - \phi$ measurements, while the TID (TEC) has four (nine) layers, providing up to four (nine) $z - \phi$ measurements. The $r - \phi$ (z) single point resolutions of these measurements are 23–35 μm (23 μm) in the TIB, 35–52 μm (52 μm) in the TOB [7]. Some layers additionally have modules mounted with stereo angles to provide measurements of the second coordinate (z in the barrel region and r in the forward regions). The single point resolutions of these measurements are $\simeq$230 μm in the TIB, $\simeq$530 μm in the TOB [8]. The hit efficiency of the silicon strip tracker is 99.8%.

2.2.4 Electromagnetic Calorimeter

Surrounding the silicon inner tracker is the electromagnetic calorimeter (ECAL), consisting of 61,200 lead tungstate ($PbWO_4$) scintillating crystals in the barrel region (EB) and 7324 crystals in each of the end-caps (EC) [7]. The EB covers out to $|\eta| < 1.479$, with inner radius 1.29 m, while the EE covers $1.479 < |\eta| < 3.0$, at a longitudinal distance $|z| = 3.15$ m from the IP (see Fig. 2.5) [7].

The ECAL measures the energy of particles, such as electrons, positrons, and photons, that primarily interact electromagnetically. Lead tungstate is high density

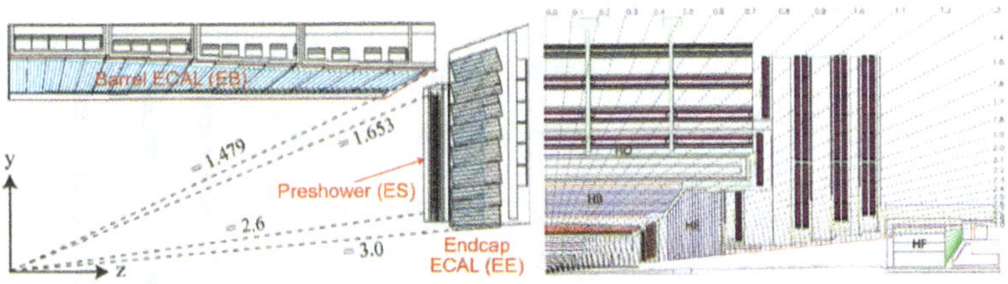

Fig. 2.5 Schematic views of the ECAL (Left) [7] and HCAL (Right) [8]

$8.28\,\mathrm{g\,cm^{-3}}$, optically transparent, radiation-hard, fast, and has a short radiation length ($X_0 = 0.89\,\mathrm{cm}$). The crystals have a highly-granular $22 \times 22\mathrm{mm}^2$ ($28.6 \times 28.6\,\mathrm{mm}^2$) front-face cross-section in the EB (EE) that matches the Molière radius of lead tungstate, i.e. the radius of a cylinder that contains a fraction 90% of the energy deposited by the shower [8]. The crystal lengths are 23 cm (22 cm) in the EB (EE), corresponding to a total radiation length of $25.8X_0$ ($24.7X_0$) [8]. The total radiation length of the crystals essentially guarantees that electrons and photons will deposit all of their energy in the ECAL, resulting in an electromagnetic shower that excites nearby atoms which subsequently relax by scintillating light. The scintillation decay time of lead tungstate is less than 25 ns, and it scintillates a blue-green light. The crystals are fully polished to totally internally reflect light down the crystal where it is collected by silicon avalanche photodiodes in the EB and vacuum phototriodes in the EE, which are designed to be radiation-hard to operate in the high magnetic field. In the $1.653 < |\eta| < 2.6$ region are pre-shower detectors that are primarily designed to distinguish photons originating from π^0 decays.

2.2.5 Hadronic Calorimeter

Outside the ECAL is the hadronic calorimeter (HCAL), consisting of the HCAL Barrel (HB), which covers out to $|\eta| < 1.3$, the HCAL End-caps (HE), which cover $1.3 < |\eta| < 3$, and radiation-hard, quartz fiber hadron calorimeters that are placed in the forward region (HF) $3 < |\eta| < 5.2$ (see Fig. 2.5) [7]. The HCAL is designed to measure jet energy by forcing their strongly-interacting hadron constituents to shower in brass absorber layers that alternate with scintillating plastic layers, in which the showers deposit their energy by exciting nearby atoms which subsequently relax by emitting light collected by hybrid photodiodes. The non-magnetic brass absorber layer is composed of 70% Cu and 30% Zn and has a nuclear interaction length of $\lambda_I = 16.42\,\mathrm{cm}$ [8]. In the barrel region, the HB does not have sufficient total interaction length for hadron showers to completely exhaust their energy in the calorimeter, so an Outer HCAL (HO) of scintillating plastic layers, that uses the solenoid as an additional absorber layer, is placed outside the solenoid and covers out to $|\eta| < 1.3$. The ECAL, which radially precedes the HCAL, contributes $1.1\lambda_I$ [8]. At $\eta = 0$, the total interaction length of the absorber layers is $5.82\lambda_I$, and increases with η, so that at $|\eta| = 1.3$ it is $10.6\lambda_I$ [8].

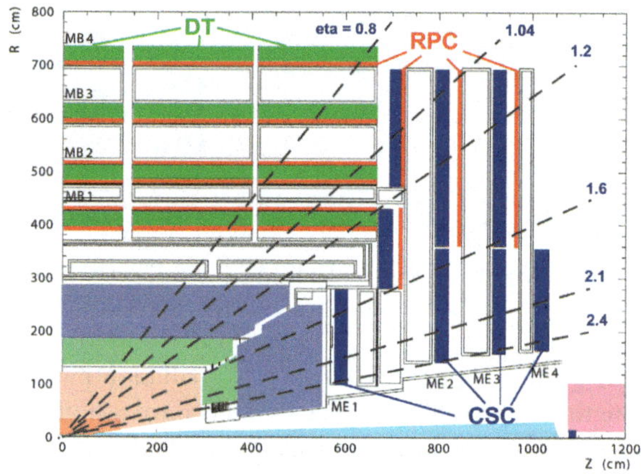

Fig. 2.6 Layout of one-quarter of the CMS muon system [7]

2.2.6 *Muon System*

Unlike most particles, muons are not stopped by any of the calorimeters, superconducting solenoid, or iron flux-return yoke. Four layers of chambers to detect muons are placed at the very outer layers of the experiment in the iron flux-return yoke. The muon system is designed with three different types of gaseous particle detectors [7]:

- drift tube (DT) chambers covering $|\eta| < 1.2$,
- cathode strip chambers (CSC) covering $0.9 < |\eta| < 2.4$, and
- resistive plate chambers (RPC) covering $|\eta| < 1.6$.

The layout for one quadrant of the muon system is shown in Fig. 2.6. The RPCs have inferior spatial resolution compared to the DTs and CSCs, but have excellent timing resolution and are primarily used for muon triggering. The muon system is aligned with the inner tracker in order for the detectors to work together to identify muons and reconstruct their trajectories.

2.2.7 *Trigger and Data Acquisition System*

CMS wants to record the most relevant data from bunch crossings with a 40 MHz event rate. Each bunch crossing generates ~1 MB of data, but due to limited bandwidth, the detectors cannot readout and store ~40 TB s^{-1}, so a trigger system is necessary to reject the least interesting 99.9999% of events in quasi-real time, and reduce the event rate to ~1 kHz before storage [8]. This is not an unreasonable constraint, as the total cross-section of proton-proton interactions (elastic and inelastic) at $\sqrt{s} = 13$ TeV is ~100 mb, while interesting processes such as top quark pair production and Higgs boson production have cross-sections on the orders of ~800 pb and ~50 pb respectively. To accomplish this, CMS has a highly reliable and efficient trigger system, composed of detector electronics, a level-1 (L1) hardware trigger, a readout network, and a high-level software trigger (HLT). The L1 trigger uses hardware processors and raw readouts from the calorimeters and muon systems to reduce the event rate to ~100 kHz with a latency of 3.8 μs [8]. The data acquisition then passes the complete readout of the event data to the HLT, which has a latency of 320 μs and performs fast reconstruction and event selection to filter events and reduce the rate of stored events to ~1 kHz [8]. The events that pass the HLT filter are saved for offline analysis to one or several HLT paths, which correspond to specific selection criteria, physics objects, or event signatures.

References

1. L. De Broglie, Recherches sur la théorie des quanta. Ph.D. dissertation, Migration- université en cours d'affectation (1924)
2. W. Heisenberg, Über den anschaulichen inhalt der quantentheoretischen kinematik und mechanik. Z. Phys. **43**(3), 172–198 (1927). ISSN: 0044-3328. [Online] Available: https://doi.org/10.1007/BF01397280
3. E. Mobs, The CERN accelerator complex. Complexe des accélérateurs du CERN (2016), General Photo. [Online]. Available: https://cds.cern.ch/record/2197559
4. L. Evans, P. Bryant, LHC machine. J. Instrum. **3**(08), S08001 (2008). [Online] Available: https://doi.org/10.1088/1748-0221/3/08/S08001
5. CMS Collaboration, Measurement of the inelastic proton-proton cross section at $\sqrt{s}$ = 13 TeV. J. High Energy Phys. **2018**(7) (2018). [Online] Available: https://doi.org/10.1007/jhep07(2018)161
6. O. Karacheban, Luminosity measurement at CMS. Presented 05 Jan 2017 (2017). [Online] Available: https://cds.cern.ch/record/2294183
7. CMS Collaboration, CMS Physics: Technical Design Report Volume 1: Detector Performance and Software (Technical design report, CMS). Geneva: CERN, 2006, There is an error on cover due to a technical problem for some items. [Online] Available: https://cds.cern.ch/record/922757
8. CMS Collaboration, The CMS experiment at the CERN LHC. The Compact Muon Solenoid experiment. JINST **3**, S08004 (2008). Also published by CERN Geneva in 2010. https://doi.org/10.1088/1748-0221/3/08/S08004. [Online] Available: http://cds.cern.ch/record/1129810
9. CMS Collaboration, Precise mapping of the magnetic field in the CMS barrel yoke using cosmic rays. JINST **5**, T03021 (2010). https://doi.org/10.1088/1748-0221/5/03/T03021. arXiv: 0910.5530. [Online] Available: http://cds.cern.ch/record/1215500
10. M. Lipinski, The phase-1 upgrade of the CMS pixel detector. J. Instrum. **12**(07), C07009 (2017). [Online] Available: https://doi.org/10.1088/1748-0221/12/07/C07009
11. CMS Collaboration, Particle-flow reconstruction and global event description with the CMS detector. Particle-flow reconstruction and global event description with the CMS detector. JINST **12**(10), P10003 (2017). Replaced with the published version. Added the journal reference and DOI. All the figures and tables can be found at http://cms-results.web.cern.ch/cms-results/public-results/publications/PRF-14-001 (CMS Public Pages). https://doi.org/10.1088/1748-0221/12/10/P10003. arXiv: 1706.04965. [Online] Available: https://cds.cern.ch/record/2270046

12. CMS Collaboration, The cms phase-1 pixel detector upgrade. J. Instrum. **16**(02), P02027 (2021). [Online] Available: https://doi.org/10.1088/1748-0221/16/02/P02027
13. CMS Collaboration, Alignment of the CMS silicon tracker during commissioning with cosmic rays. JINST **5**, T03009 (2010). https://doi.org/10.1088/1748-0221/5/03/T03009. arXiv: 0910.2505. [Online] Available: https://cds.cern.ch/record/1211825

Chapter 3
Top Quark Physics at the Large Hadron Collider

Abstract This chapter focuses on the relevant properties and physics of top quarks for understanding the results of this dissertation. The production and decay mechanisms of top quarks at hadronic colliders are summarized, with particular emphasis the spin properties of $t\bar{t}$ production and top quark decay. Context is provided by reviewing previous measurements of top quark polarization and $t\bar{t}$ spin correlations, setting a context for the analyses presented in this dissertation. The significance of top quark studies for precision measurements and potential new physics discoveries is provided, emphasizing the top quark as a probe for both SM and BSM physics.

The top quark was predicted in 1973 by Kobayashi and Maskawa when the Cabibbo matrix was extended to the CKM matrix [1] in order to accommodate the experimental discovery of CP-violation in precision measurements of kaon regeneration [2]. This extension necessarily required three quark generations, when at the time only the up, down, and strange quarks were known. The charm quark was discovered in 1974 [3, 4] and the bottom quark in 1977 [5]. After searching for two decades, the top quark was finally discovered in 1995 at Fermi National Acceleration Laboratory (FNAL), jointly by the CDF [6] and D0 [7] Tevatron Experiments. It is the heaviest elementary particle ever discovered, with a measured mass $m_t = 172.69 \pm 0.30$ GeV [8], the top quark has a mass equivalent to a Tungsten isotope containing 184 nucleons; it is $\sim$35 times heavier than the next heaviest quark, the bottom quark, as illustrated in Fig. 3.1. Due to its very large mass, the top quark has a very short lifetime, or equivalently, a large decay width. The top quark full decay width is larger than the QCD hadronization scale and much larger than the spin decorrelation scale. Thus, not only do top quarks decay before hadronizing[1] to form bound states but also their properties, including spin information, are transferred to their decay products. The spin information of top quarks is observable

[1] Theoretical predictions for toponium bound state production at the $\sqrt{s} = 13$ TeV LHC are around 0.8% the total $t\bar{t}$ production cross-section [10]. Experimental evidence for toponium production has not yet been observed.

J. R. Thieman, *Top Quark Spin Correlations with the CMS Detector*,
Springer Theses, https://doi.org/10.1007/978-3-031-76117-1_3

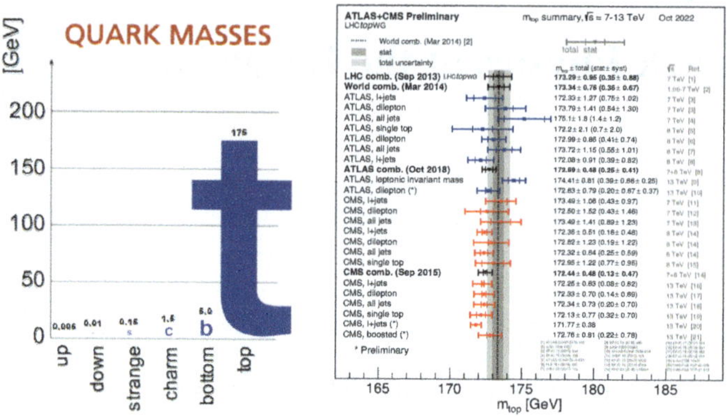

Fig. 3.1 The top quark is the heaviest elementary particle ever discovered and is ~35 times heavier than the next heaviest quark (Left). Summary of the ATLAS and CMS direct measurements of top quark mass (Right) [9]

in the angular kinematic distributions of their decay products. These properties make top quarks unique in the quark sector, as top quark measurements are the only opportunity to study the properties of bare quarks, free of QCD confinement and hadronization effects.

3.1 Top Quark Pair Production

Because of the large mass, the production of top quarks has only been possible by Tevatron and LHC high-energy accelerator experiments. At these colliders, the majority of top quarks are produced in $t\bar{t}$ pairs by the strong interaction, but single t production by the electroweak interaction is also possible. For $t\bar{t}$ production, three parton production processes are possible at hadron colliders: gluon-gluon (gg) fusion, quark/anti-quark ($q\bar{q}$) annihilation, and quark-gluon scattering. While at the Tevatron $q\bar{q}$ annihilation was the dominant production mechanism, for $\sqrt{s} = 13\,\text{TeV}$ proton-proton collisions at the LHC, the gluon density of the proton parton distribution function (PDF) is much higher than the quark density (see Sect. 4.1), and gg fusion accounts for ~90% of $t\bar{t}$ production.

The inclusive $t\bar{t}$ production cross-section in proton-proton (pp) collisions is calculated by convoluting the proton PDFs with the partonic cross-sections and summing over all parton constituents of the proton, i.e. gluons, valence quarks, and sea quarks. For QCD $t\bar{t}$ production at leading-order (LO), three channels of gg fusion and one channel of $q\bar{q}$ annihilation parton production processes are possible. LO Feynman diagrams for these processes are shown in Fig. 3.2. Quark-gluon scattering can also produce $t\bar{t}$ pairs, but only at next-to-leading-order (NLO) and higher orders. At next-to-next-to-leading-order (NNLO) with

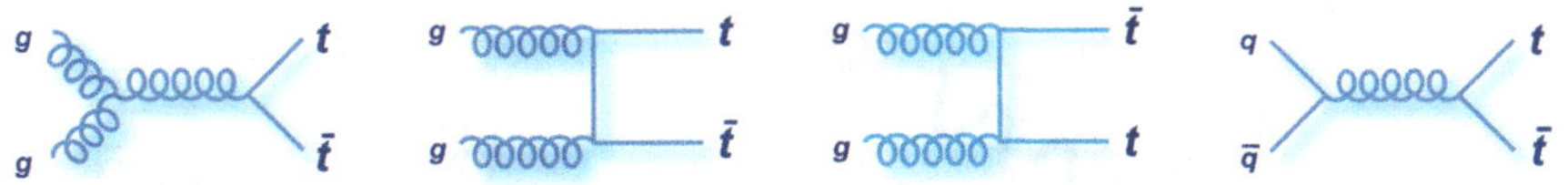

Fig. 3.2 LO Feynman diagrams for QCD $t\bar{t}$ production: three channels of gg fusion and one channel of $q\bar{q}$ annihilation [11]

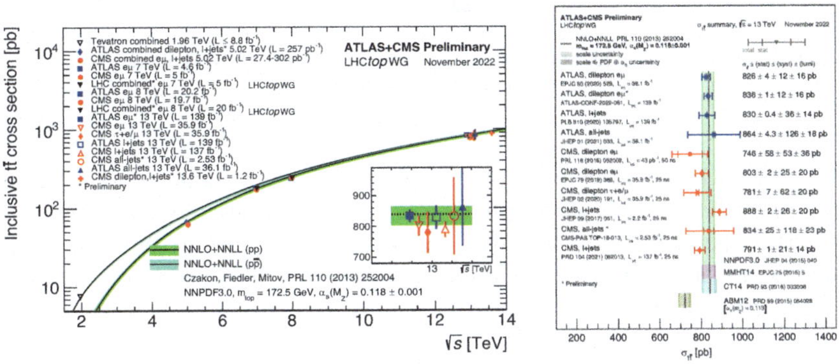

Fig. 3.3 Summary of LHC and Tevatron measurements of the top-pair production cross-section as a function of the center-of-mass energy compared to the NNLO QCD calculation complemented with NNLL resummation (Left). Summary of measurements of the top-pair production cross-section at $\sqrt{s} = 13$ TeV compared to the exact NNLO QCD calculation complemented with NNLL resummation (Right) [9]

next-to-next-to-leading-logarithmic (NNLL) precision for $\sqrt{s} = 13\,\text{TeV}$, assuming $m_t = 172.5\,\text{GeV}$, the inclusive $t\bar{t}$ production cross-section is calculated to be $\sigma_{t\bar{t}} = 831.76^{+19.77}_{-29.20}$ (Scale) ± 35.06 (PDF + α_S) pb. A summary of $t\bar{t}$ cross-section measurements by LHC and Tevatron experiments as a function of the center-of-mass energy, as well as a summary of $t\bar{t}$ cross-section measurements by ATLAS and CMS at $\sqrt{s} = 13$ TeV are shown in Fig. 3.3.

3.2 Decays of Top Quark Pairs

The top quark decay width Γ_t implies a lifetime of $\tau_t = 1/\Gamma_t = 4.6 \times 10^{-25}$ s before decaying via the weak interaction into an on-shell W-boson and down-type quark ($t \to W^+ q$). The branching ratio of each mode can be calculated using elements of the CKM matrix $B_x = \frac{|V_{tx}|^2}{|V_{tb}|^2 + |V_{ts}|^2 + |V_{td}|^2}$, and while any of the down-type quarks are possible, because $|V_{tb}| = 0.999118 >> |V_{ts}| = 0.04110$ or $|V_{td}| = 0.00857$ [8], $t \to W^+ b$ dominates with a branching ratio $B_b \simeq 99.8\%$.

The W-boson can decay hadronically $W^+ \to q\bar{q}'$, where q is an up-type quark (u or c) and $\bar{q}'$ is an anti-down-type quark ($\bar{d}$, $\bar{s}$, or $\bar{b}$), or leptonically

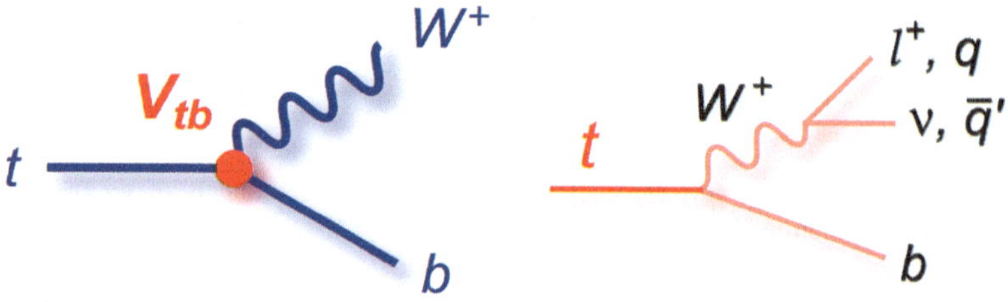

Fig. 3.4 The branching modes of $t \to W^+q$ depend on the values of CKM matrix elements (Left). The W-boson can decay hadronically or leptonically (Right) [11]

$W^+ \to \bar{\ell}\nu_\ell$, where $\bar{\ell}$ is an anti-lepton ($\bar{e}$, $\bar{\mu}$, or $\bar{\tau}$) and ν_ℓ is a neutrino of the same lepton flavor (Fig. 3.4). Ignoring phase space considerations, the relative branching fractions of the hadronic and leptonic decay modes are given by the ratio of flavor combinations and color combinations, together with the corresponding squared CKM matrix elements:

$$\begin{aligned}\frac{\text{hadronic}}{\text{leptonic}} &= \frac{\{u\bar{d}:|V_{ud}|^2, u\bar{s}:|V_{us}|^2, u\bar{b}:|V_{ub}|^2, c\bar{d}:|V_{cd}|^2, c\bar{s}:|V_{cs}|^2, c\bar{b}:|V_{cb}|^2\}\times\{R,B,G\}}{\{e\bar{\nu}_e, \mu\bar{\nu}_\mu, \tau\bar{\nu}_\tau\}} \\ &= \frac{6}{3} \end{aligned} \tag{3.1}$$

where decays with top quarks are forbidden due to energy conservation, and due to the unitary of the CKM matrix $|V_{ud}|^2+|V_{us}|^2+|V_{ub}|^2 = |V_{cd}|^2+|V_{cs}|^2+|V_{cb}|^2 = 1$. Including phase space considerations, the hadronic W-boson decay mode occurs $(66.5 \pm 1.4)\%$ of the time, and the leptonic mode occurs $(33.2 \pm 1.0)\%$ [8].

For a $t\bar{t}$ pair, depending on the W-boson decays of the t and $\bar{t}$, there are three decay modes:

- **Fully Hadronic:** $t\bar{t} \to W^+b\ W^-\bar{b} \to q\bar{q}'b\ \bar{q}''q'''\bar{b}$, also referred to as "all jets"
- **Semi-Leptonic:** $t\bar{t} \to W^+b\ W^-\bar{b} \to q\bar{q}'b\ \ell\bar{\nu}_\ell\bar{b}$ or $t\bar{t} \to W^+b\ W^-\bar{b} \to \bar{\ell}\nu_\ell b\ \bar{q}''q'''\bar{b}$, also referred to as "lepton + jets"
- **Dileptonic:** $t\bar{t} \to W^+b\ W^-\bar{b} \to \bar{\ell}\nu_\ell b\ \ell'\bar{\nu}_{\ell'}\bar{b}$

Examples of LO Feynman diagrams for the $t\bar{t}$ decay modes are shown in Fig. 3.5. Because hadronic W-boson decay modes are more common than leptonic ones, for $t\bar{t}$ decays the dileptonic mode occurs least often ~10.5% [8]. A schematic classifying $t\bar{t}$ decay modes, together with a chart showing the relative branching fractions, is shown in Fig. 3.6.

Furthermore, τ can decay either hadronically or leptonically via a virtual W-boson. For lepton decays of $\tau \to \ell\bar{\nu}_\ell\nu_\tau$, the τ decays into a lepton, an anti-neutrino of the same lepton flavor, and a τ flavor neutrino. The lepton branching fractions

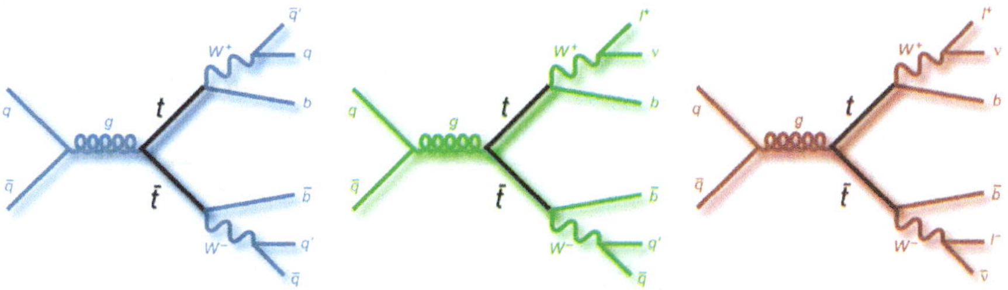

Fig. 3.5 Examples of LO Feynman diagrams for the $t\bar{t}$ decay modes: Fully Hadronic (Left), Semi-Leptonic (Middle), and Dileptonic (Right) [11]

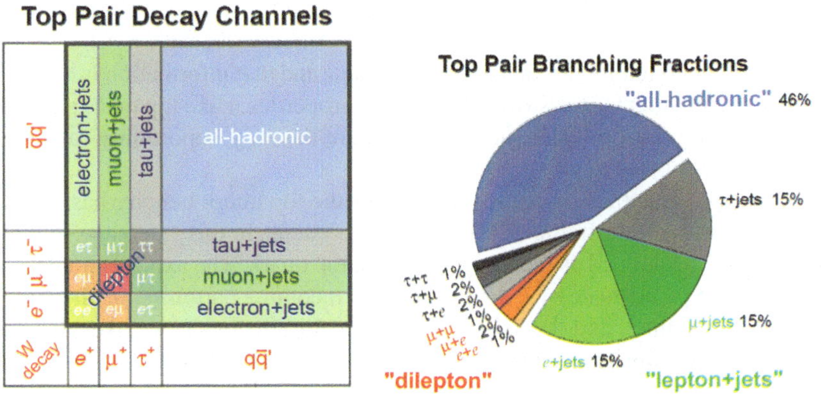

Fig. 3.6 A schematic classifying $t\bar{t}$ decay modes (Left), and chart showing the relative branching fractions (Right) [11]

of τ decays are $\sim(17.82 \pm 0.04)\%$ for $\ell = e$ and $(17.39 \pm 0.04)\%$ for $\ell = \mu$ [8]. Including τ decays, the $t\bar{t}$ dileptonic final state branching fraction is $\sim 6.42\% = \sim 4.55\%$ (prompt)$+$ $\sim 1.87\%$ (via tau). The $t\bar{t}$ dileptonic final state is further categorized into three channels: e^+e^-, $e^\pm\mu^\mp$, and $\mu^+\mu^-$, with the combination of the three channels, denoted $\ell\bar{\ell}$.

3.3 Top Quark Spin in $t\bar{t}$ Production and Top Decay

In general, hadronization of quarks dilutes their spin information. Moreover, the spins of quarks can be depolarized by radiative processes such as gluon emission, causing spin decorrelation. For the top quark however, it has already been mentioned that the full decay width $\Gamma_t = 1.42^{+0.19}_{-0.15}$ GeV [8] is larger than the QCD hadronization scale ($\Lambda_{QCD} \simeq 0.38$ GeV [12]) and much larger than the spin decorrelation scale ($\Lambda^2_{QCD}/m_t \sim 0.08$ MeV [13]). Thus, not only does a top quark decay before hadronizing to form bound states but also its spin properties are

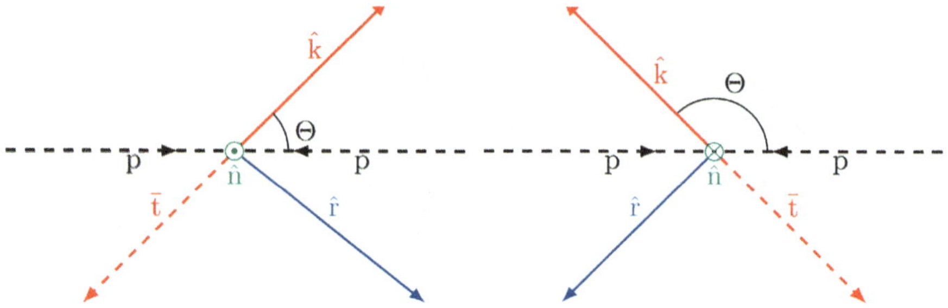

Fig. 3.7 The helicity basis spin quantization axes in the $t\bar{t}$ ZMF. $\hat{n}$ and $\hat{r}$ flip signs as Θ crosses $\frac{\pi}{2}$

transferred, without dilution, to its decay products and that information is observable in their angular kinematic distributions. These properties make top quarks unique in the quark sector, as top quark measurements are the only opportunity to study the spin properties of free quarks.

For the spin quantization axes, we consider the top quark helicity basis $(\hat{k}, \hat{n}, \hat{r})$ in the zero momentum frame (ZMF) of the $t\bar{t}$ system. In this reference frame, the helicity axis $\hat{k}$, is defined as the direction of the top quark. The axis $\hat{n}$ is transverse to the production plane made by $\hat{k}$ direction of the incoming parton $\hat{p}$, i.e. the direction of the proton beam $\hat{z}$, and the axis $\hat{r}$ is orthogonal to the other two axes, forming a right-handed coordinate system:

$$\begin{aligned} \hat{n} &= \frac{\text{sign}(\cos\Theta)}{\sin\Theta}(\hat{p} \times \hat{k}) \\ \hat{r} &= \frac{\text{sign}(\cos\Theta)}{\sin\Theta}(\hat{p} - \hat{k}\cos\Theta) \end{aligned} \tag{3.2}$$

where Θ is the top quark scattering angle and sign$(\cos\Theta)$ defines a "forward" direction for each event, the importance of which will be made explicit later, so that $\hat{n}$ and $\hat{r}$ flip signs as Θ crosses $\frac{\pi}{2}$. The helicity basis spin quantization axes are shown in Fig. 3.7.

At LO, the kinematic and spin properties of a $t\bar{t}$ pair can be completely described in the $t\bar{t}$ ZMF by the invariant mass $m_{t\bar{t}}$ and the top scattering angle Θ. Helicity is defined as spin direction projected onto the linear momentum direction, and for massive particles, helicity is reference frame dependent. In the helicity basis of the $t\bar{t}$ ZMF, squared matrix elements (see Sect. 4.2) for the $q\bar{q}$ annihilation LO Feynman diagram in Fig. 3.2, separated into like-helicity (LL, RR) and unlike-helicity (LR, RL) $t\bar{t}$ pairs, are [14]:

$$\begin{aligned} \textstyle\sum_{LL,RR} |\mathcal{M}(q\bar{q} \to t\bar{t})|^2 &= 8(4\pi\alpha_S)^2 \left(1 - \beta^2\right) \sin^2\Theta \\ \textstyle\sum_{LR,RL} |\mathcal{M}(q\bar{q} \to t\bar{t})|^2 &= 8(4\pi\alpha_S)^2 \left(1 + \cos^2\Theta\right) \end{aligned} \tag{3.3}$$

where $\beta = \sqrt{1 - 4m_t^2/m_{t\bar{t}}^2}$ is the speed of the top quark (as a fraction of c) in the $t\bar{t}$ ZMF and is completely determined by $m_{t\bar{t}}$. In the helicity basis of the $t\bar{t}$ ZMF, squared matrix elements for the gg fusion LO Feynman diagrams in Fig. 3.2 separated into like-helicity (LL, RR) and unlike-helicity (LR, RL) $t\bar{t}$ pairs are [14]:

$$\sum\nolimits_{LL,RR} |\mathcal{M}(gg \to t\bar{t})|^2$$
$$= \frac{16}{3}(4\pi\alpha_S)^2\left(\frac{7+9\beta^2\cos^2\Theta}{(1-\beta^2\cos^2\Theta)^2}\right)\left(1-\beta^2\right)\left(1+\beta^2+\beta^2\sin^4\Theta\right)$$
$$\sum\nolimits_{LR,RL} |\mathcal{M}(gg\to t\bar{t})|^2 = \frac{16}{3}(4\pi\alpha_S)^2\left(\frac{7+9\beta^2\cos^2\Theta}{(1-\beta^2\cos^2\Theta)^2}\right)\beta^2\sin^2\Theta\left(1+\cos^2\Theta\right). \tag{3.4}$$

These expressions include summation over the spins of the incoming partons, as well as summation over the colors of both the incoming and outgoing states, but factors averaging over the spins and colors have not been included.

Equations 3.3 indicate that in the high energy limit ($\beta \to 1$), the production of like-helicity $t\bar{t}$ pairs by $q\bar{q}$ annihilation is suppressed. Equations 3.4 indicate that in the high energy limit ($\beta \to 1$), the production of like-helicity $t\bar{t}$ pairs by gg fusion is also suppressed, but additionally at low energies $\beta << 1$, production of unlike-helicity $t\bar{t}$ pairs is suppressed relative to the production of like-helicity pairs by a factor (β^2). Figure 3.8 shows the breakdown of the total $t\bar{t}$ cross-section into like-helicity and unlike-helicity pairs as a function of $m_{t\bar{t}}$ using the helicity basis for the $\sqrt{s} = 14$ TeV LHC. Near threshold, the dominant spin states of the $t\bar{t}$ system are $^1S_0$ for like-helicity $t\bar{t}$ pairs from gg fusion and $^3S_1$ for unlike-helicity $t\bar{t}$ pairs from $q\bar{q}$ annihilation:

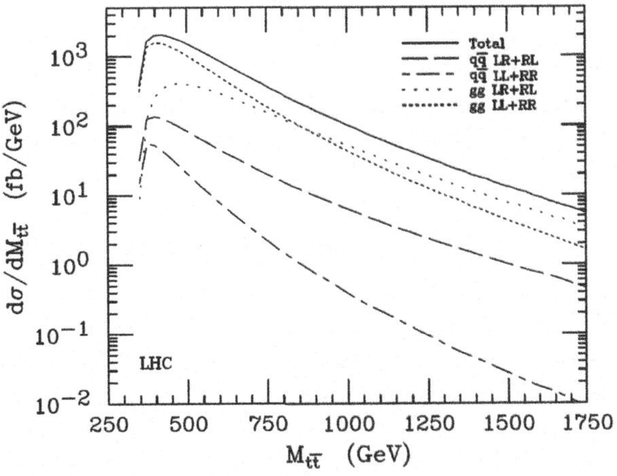

Fig. 3.8 Differential cross section for $t\bar{t}$ production as a function of $m_{t\bar{t}}$, for the $\sqrt{s} = 14$ TeV LHC, decomposed into $LL + RR$ and $LR + RL$ helicities in the zero momentum frame of the $t\bar{t}$ pair for both $q\bar{q}$ and gg components using the helicity basis [14]

Table 3.1 The top spin analyzing power at NLO for the various daughter and granddaughter decay particles of the top quark [15]

Top spin analyzing power at NLO	
κ_b	−0.39
κ_{W^+}	0.39
$\kappa_{\bar{\ell}}$	**0.998**
$\kappa_{\bar{d},\bar{s}}$	0.93
$\kappa_{u,c}$	−0.31

The bold value indicates the mode used in this analysis.

$$^1S_0 : \left\{ \frac{1}{\sqrt{2}}[|\uparrow\downarrow\rangle - |\downarrow\uparrow\rangle] \quad ^3S_1 : \begin{cases} |\uparrow\uparrow\rangle \\ \frac{1}{\sqrt{2}}[|\uparrow\downarrow\rangle + |\downarrow\uparrow\rangle] \\ |\downarrow\downarrow\rangle \end{cases} \right. \tag{3.5}$$

where the first (second) arrow of $|\uparrow\downarrow\rangle$ refers to the spin state of the top quark (anti-quark) along spin quantization axis $\hat{i}$ ($\hat{j}$). It will be demonstrated in Sect. 3.4 that the different parton production processes and helicity combinations manifest themselves in the $t\bar{t}$ final state particle angular kinematic distributions.

The differential decay rate of top quark daughter or granddaughter particle x, for an ensemble of polarized top quarks at rest with polarization $0 \leq \vec{P} \leq 1$, can be parameterized [15] as:

$$\frac{1}{\Gamma}\frac{d\Gamma}{d(\cos\theta_x)} = \frac{1}{2}[1 + |\vec{P}|\kappa_x \cos\theta_x] \tag{3.6}$$

where θ_x is the angle between $\vec{P}$ and the momentum direction of x in the top quark ZMF and κ_x is the top spin analyzing power of particle x, denoting what fraction of the top polarization information gets transferred from the top to its daughter or granddaughter particle x. The top spin analyzing power at NLO is shown for the various daughter and granddaughter particles in Table 3.1. For charged lepton granddaughters of top quarks, the top spin analyzing power (κ_ℓ) is close to unity, and when combined with the experimental advantages that charged leptons can be triggered on efficiently and reconstructed precisely, it is concluded that the leptonic decay mode is the optimal proxy for studies of top quark spin properties.

3.4 Top Quark Polarizations and t$\bar{\text{t}}$ Spin Correlations

Top quarks and anti-quarks, produced in $t\bar{t}$ pairs at hadron colliders via the strong interaction, are not polarized at LO due to the parity conserving nature of QCD (longitudinal polarization) and approximate T-invariance of QCD (transverse polarization). However, as was shown in the previous section, the spins of top quarks produced in $t\bar{t}$ pairs are strongly correlated with spins of anti-top quarks, with dependence on $m_{t\bar{t}}$, Θ, and choice of spin quantization axes. The squared matrix element for the $t\bar{t}$ production and decay can be written as [16]:

$$|\mathcal{M}(gg/q\bar{q} \rightarrow t\bar{t} \rightarrow (\bar{\ell}\nu_\ell b)(\ell'\bar{\nu_{\ell'}}\bar{b}))|^2 \sim Tr[\rho R\bar{\rho}] \tag{3.7}$$

where R is the spin density matrix related to on-shell $t\bar{t}$ production and ρ is the decay density matrix. The production spin density matrix R can be decomposed in the spin spaces of t and $\bar{t}$ using a basis of Pauli matrices [16]:

$$R^I \propto \tilde{A}\mathbb{1} \otimes \mathbb{1} + \tilde{B}_i^+\sigma^i \otimes \mathbb{1} + \tilde{B}_j^-\mathbb{1} \otimes \sigma^j + \tilde{C}_{ij}\sigma^i \otimes \sigma^j \tag{3.8}$$

where initial state $I = gg/q\bar{q}$ at fixed energy and top direction, $\tilde{A}$ determines the differential cross section for $t\bar{t}$ production, $\mathbb{1}$ is the 2×2 identity matrix, $\sigma^{i(j)}$ are the Pauli matrices, $\tilde{B}^\pm$ are 3-vectors that characterize the top quark $(+)$ and anti-quark $(-)$ polarizations along each of the spin quantization axes, and $\tilde{C}$ is a 3×3 matrix that characterizes the correlations between the top quark and anti-quark spins along each pair of spin quantization axes. $\tilde{B}_i^\pm$ and $\tilde{C}_{ij}$ can be further decomposed using the $t\bar{t}$ ZMF helicity basis $\{\hat{k}, \hat{n}, \hat{r}\}$ [16]:

$$\begin{aligned}
\tilde{B}_i^\pm &= b_k^\pm \hat{k}_i + b_n^\pm \hat{n}_i + b_r^\pm \hat{r}_i \\
\tilde{C}_{ij} &= c_{kk}\hat{k}_i\hat{k}_j + c_{nn}\hat{n}_i\hat{n}_j + c_{rr}\hat{r}_i\hat{r}_j \\
&\quad + c_{rk}(\hat{r}_i\hat{k}_j + \hat{k}_i\hat{r}_j) + c_{kn}(\hat{k}_i\hat{n}_j + \hat{n}_i\hat{k}_j) + c_{nr}(\hat{n}_i\hat{r}_j + \hat{r}_i\hat{n}_j) \\
&\quad + c_n(\hat{r}_i\hat{k}_j - \hat{k}_i\hat{r}_j) + c_r(\hat{k}_i\hat{n}_j - \hat{n}_i\hat{k}_j) + c_k(\hat{n}_i\hat{r}_j - \hat{r}_i\hat{n}_j)
\end{aligned} \tag{3.9}$$

where the coefficients $b^\pm$ and c are functions of $m_{t\bar{t}}$ and $\cos\Theta$. The helicity basis is chosen because of its stability with regard to reconstruction errors and the spin correlations are strong using this basis. By construction, the coefficient functions $b^\pm$ and c have definite properties with respect to discrete symmetries P, CP, T_N (naive time-reversal symmetry at tree-level), and identical particle exchange ($Bose$), and as a result, these coefficients can help distinguish NP effects with SMEFT [16]. The P, CP, and T invariance of QCD forces many of these coefficients to be zero before considering electroweak and higher-order QCD loop corrections. The Bose symmetry of $I = gg$ implies $R^{gg}(\hat{k}, \hat{p}) = R^{gg}(-\hat{k}, \hat{p})$; hence the importance of $\text{sign}(\cos\Theta)$ in the definition of $\{\hat{k}, \hat{n}, \hat{r}\}$ (see Eq. 3.2), to allow non-zero coefficients when the parton production process is gg fusion. A summary of transformation properties of the coefficient functions with respect to P, CP, T_N, and $Bose$ ($I = gg$ only) symmetries [16] is shown in Table 3.2.

As has been already mentioned, top quark spins cannot be measured directly but can be traced in the differential angular distributions of their charged lepton granddaughters. For $t\bar{t}$ dileptonic events, the normalized fourfold angular cross-section [16] for the two leptons is:

$$\frac{1}{\sigma}\frac{d\sigma}{d\Omega_1 d\Omega_2} = \frac{1}{(4\pi)^2}\left(1 + \vec{B}_1 \cdot \hat{\ell}_1 + \vec{B}_2 \cdot \hat{\ell}_2 - \hat{\ell}_1 \cdot \mathbf{C} \cdot \hat{\ell}_2\right) \tag{3.10}$$

Table 3.2 Transformation properties of the coefficient functions with respect to CP, P, and T_N symmetries for $I = gg, q\bar{q}$ [16]. In the last column, all the coefficients of the initial state $I = gg$ are even under Bose symmetry because of the sign($\cos\Theta$) factor in the definition of $\{\hat{k}, \hat{n}, \hat{r}\}$ (see Eq. 3.2)

	CP	P	T_N	CPT_N	Bose
A^I	A^I	A^I	A^I	A^I	A^{gg}
$b_r^{I\pm}$	$b_r^{I\mp}$	$-b_r^{I\pm}$	$b_r^{I\pm}$	$b_r^{I\mp}$	$b_r^{gg\pm}$
$b_k^{I\pm}$	$b_k^{I\mp}$	$-b_k^{I\pm}$	$b_k^{I\pm}$	$b_k^{I\mp}$	$b_k^{gg\pm}$
$b_n^{I\pm}$	$b_n^{I\mp}$	$b_n^{I\pm}$	$-b_n^{I\pm}$	$-b_n^{I\mp}$	$b_n^{gg\pm}$
c_{rr}^I	c_{rr}^I	c_{rr}^I	c_{rr}^I	c_{rr}^I	c_{rr}^{gg}
c_{kk}^I	c_{kk}^I	c_{kk}^I	c_{kk}^I	c_{kk}^I	c_{kk}^{gg}
c_{nn}^I	c_{nn}^I	c_{nn}^I	c_{nn}^I	c_{nn}^I	c_{nn}^{gg}
c_{rk}^I	c_{rk}^I	c_{rk}^I	c_{rk}^I	c_{rk}^I	c_{rk}^{gg}
c_{rn}^I	c_{rn}^I	$-c_{rn}^I$	$-c_{rn}^I$	$-c_{rn}^I$	c_{rn}^{gg}
c_{kn}^I	c_{kn}^I	$-c_{kn}^I$	$-c_{kn}^I$	$-c_{kn}^I$	c_{kn}^{gg}
c_r^I	$-c_r^I$	$-c_r^I$	$-c_r^I$	c_r^I	c_r^{gg}
c_k^I	$-c_k^I$	$-c_k^I$	$-c_k^I$	c_k^I	c_k^{gg}
c_n^I	$-c_n^I$	c_n^I	c_n^I	$-c_n^I$	c_n^{gg}

where $\hat{\ell}_1$ ($\hat{\ell}_2$) is the charged lepton direction in the ZMF of its grandmother top quark (anti-quark). $\vec{B}_{1(2)}$ are 3-vectors sensitive to $\tilde{B}^{\pm}$, and **C** is a 3×3 matrix sensitive to $\tilde{C}$. Each coefficient probes a single coefficient function:

- The top (anti-top) polarization coefficients are B_1^i (B_2^j), with respect to spin quantization axis $\hat{i}$ ($\hat{j}$), and are sensitive to coefficient functions b_i^+ (b_j^-).
- The diagonal spin correlation coefficients are C_{ii}, with respect to spin quantization axis $\hat{i}$, and are sensitive to c_{ii}.
- The cross spin correlation coefficients are C_{ij} and C_{ji}, for each pair of spin quantization axes $\hat{i}$ and $\hat{j}$, with the sums and differences $C_{ij} \pm C_{ji}$ sensitive to c_{ij} and c_i 3.9.

Unlike the coefficient functions of $\tilde{B}^{\pm}$ and $\tilde{C}$, the coefficients of $\vec{B}_{1(2)}$ and **C** are directly observable final state variables. A summary of the polarization and spin correlation coefficients, which coefficient functions they probe, and which contributions in the differential parton cross-sections the coefficients are sensitive to, is summarized in Table 3.3.

Integrating out the azimuthal angles of the normalized fourfold angular cross-section 3.10 yields the normalized polar angle double cross-section,

Table 3.3 A summary of the polarization and spin correlation coefficients, which coefficient functions they probe, and which contributions in the differential parton cross-sections the coefficients are sensitive to [16]

Coefficients	Coefficient functions	Contributions
C_{nn}	c^I_{nn}	P-, CP-even
C_{rr}	c^I_{rr}	P-, CP-even
C_{kk}	c^I_{kk}	P-, CP-even
$C_{rk} + C_{kr}$	c^I_{rk}	P-, CP-even
$C_{nr} + C_{rn}$	c^I_{rn}	P-odd, CP-even, absorptive
$C_{nk} + C_{kn}$	c^I_{kn}	P-odd, CP-even, absorptive
$C_{rk} - C_{kr}$	c^I_n	P-even, CP-odd, absorptive
$C_{nr} - C_{rn}$	c^I_k	P-odd, CP-odd
$C_{nk} - C_{kn}$	$-c^I_r$	P-odd, CP-odd
$B^n_1 + B^n_2$	$b^{I+}_n + b^{I-}_n$	P-, CP-even, absorptive
$B^n_1 - B^n_2$	$b^{I+}_n - b^{I-}_n$	P-even, CP-odd
$B^r_1 + B^r_2$	$b^{I+}_r + b^{I-}_r$	P-odd, CP-even
$B^r_1 - B^r_2$	$b^{I+}_r - b^{I-}_r$	P-odd, CP-odd, absorptive
$B^k_1 + B^k_2$	$b^{I+}_k + b^{I-}_k$	P-odd,CP-even
$B^k_1 - B^k_2$	$b^{I+}_k - b^{I-}_k$	P-odd, CP-odd, absorptive

$$\frac{1}{\sigma}\frac{d\sigma}{d\cos\theta^i_1 d\cos\theta^j_2} = \frac{1}{4}\left(1 + B^i_1\cos\theta^i_1 + B^j_2\cos\theta^j_2 - C_{ij}\cos\theta^i_1\cos\theta^j_2\right), \quad \text{where} \tag{3.11}$$

$$\begin{aligned} B^i_1 &= P_i\kappa_\ell, \\ B^j_2 &= -\bar{P}_j\kappa_\ell, \\ C_{ij} &= \kappa^2_\ell \frac{\sigma(\uparrow\uparrow)+\sigma(\downarrow\downarrow)-\sigma(\uparrow\downarrow)-\sigma(\downarrow\uparrow)}{\sigma(\uparrow\uparrow)+\sigma(\downarrow\downarrow)+\sigma(\uparrow\downarrow)+\sigma(\downarrow\uparrow)} \end{aligned} \tag{3.12}$$

with $P_i = \langle 2\vec{S}_t \cdot \hat{i}\rangle$, $\bar{P}_j = \langle 2\vec{S}\bar{t} \cdot \hat{j}\rangle$, and the first (second) arrow of $\sigma(\uparrow\downarrow)$ referring to the spin state of the top quark (anti-quark) along spin quantization axis $\hat{i}$ ($\hat{j}$). The quantity $\cos\theta^k_1 = \hat{\ell}_1 \cdot \hat{k}$, as an example, can be calculated by taking the dot product of the charged lepton direction in its grandfather top quark's ZMF with the $\hat{k}$ helicity basis spin quantization axis in the $t\bar{t}$ ZMF and can be demonstrated to be Lorentz invariant.

From Eq. 3.11, it can be derived that the single-differential cross-sections are given by:

$$\frac{1}{\sigma}\frac{d\sigma}{d\cos\theta^i_{1(2)}} = \frac{1}{2}\left(1 + B^i_{1(2)}\cos\theta^i_{1(2)}\right), \tag{3.13}$$

$$\frac{1}{\sigma}\frac{d\sigma}{d\cos\theta^i_1\cos\theta^i_2} = \frac{1}{2}\left(1 - C_{ii}\cos\theta^i_1\cos\theta^i_2\right)\log\left(\frac{1}{|\cos\theta^i_1\cos\theta^i_2|}\right), \tag{3.14}$$

$$\frac{1}{\sigma}\frac{\mathrm{d}\sigma}{\mathrm{d}x_\pm} = \frac{1}{2}\left(1 - \frac{C_{ij}\pm C_{ji}}{2}x_\pm\right)\cos^{-1}|x_\pm| \tag{3.15}$$

for $\hat{i} \neq \hat{j}$ where $x_\pm = \cos\theta^i_1\cos\theta^j_2 \pm \cos\theta^j_1\cos\theta^i_2$.

These single-differential cross-sections would otherwise be symmetric functions if not for the presence of the coefficient parameters. Thus the fifteen coefficients can be extracted from the mean or asymmetry of the following single-differential cross-sections:

- The six top quark (anti-quark) polarization coefficients $B^i_{1(2)}$ using 3.13,
- The three diagonal spin correlation coefficients C_{ii} using 3.14, and
- Sums and differences of the six cross spin correlation coefficients $C_{ij} \pm C_{ji}$ using 3.15.

The above decomposition is true for any subset of the full phase space, so double differential measurements of the polarization and spin correlations involve partitioning the phase space and, in each partition, repeating the procedure of measuring the relevant single-differential cross-sections and then extracting the coefficients.

3.5 Previous Measurements of Top Quark Polarizations and t$\bar{\text{t}}$ Spin Correlations at the LHC

Measurements of $t\bar{t}$ spin correlations in the dilepton mode have been performed by both the Compact Muon Solenoid (CMS) and ATLAS experiments using 36 fb^{-1} of data taken at the LHC during 2016 at a center-of-mass collision energy of $\sqrt{s} = 13$ TeV [17–19]. Both experiments unfolded their results to parton-level and extrapolated to the full phase space. Both CMS and ATLAS made indirect measurements of $t\bar{t}$ spin correlations by measuring $|\Delta\phi_{\ell\bar{\ell}}|$, the difference in azimuthal angle between the two leptons in the laboratory frame, but ATLAS additionally measured $|\Delta\eta_{\ell\bar{\ell}}|$, the difference in pseudorapidity between the two leptons in the laboratory frame, and performed their $|\Delta\phi_{\ell\bar{\ell}}|$ measurements differentially as a function of $t\bar{t}$

invariant mass, but only made measurements in the $e^{\pm}\mu^{\mp}$ channel. ATLAS observed a 3.2σ deviation from the SM in their measurement of $|\Delta\phi_{\ell\bar{\ell}}|$ remains in tension with the CMS result, but could possibly be explained by missing higher-order corrections to the top quark kinematics in Monte Carlo (MC) simulations.

References

1. M. Kobayashi, T. Maskawa, CP-violation in the renormalizable theory of weak interaction. Prog. Theor. Phys. **49**(2), 652–657 (1973). ISSN: 0033-068X. eprint: https://academic.oup.com/ptp/article-pdf/49/2/652/5257692/49-2-652.pdf. [Online] Available: https://doi.org/10.1143/PTP.49.652
2. J.H. Christenson, J.W. Cronin, V.L. Fitch, R. Turlay, Evidence for the 2π decay of the K_2^0 meson. Phys. Rev. Lett. **13**, 138–140 (1964). [Online]. Available: https://doi.org/10.1103/PhysRevLett.13.138
3. J.J. Aubert, U. Becker, P.J. Biggs, et al., Experimental observation of a heavy particle J. Phys. Rev. Lett. **33**, 1404–1406 (1974). [Online]. Available: https://doi.org/10.1103/PhysRevLett.33.1404
4. J.E. Augustin, A.M. Boyarski, M. Breidenbach, et al., Discovery of a narrow resonance in e^+e^- annihilation. Phys. Rev. Lett. **33**, 1406–1408 (1974). [Online]. Available: https://doi.org/10.1103/PhysRevLett.33.1406
5. S.W. Herb, D.C. Hom, L.M. Lederman, et al., Observation of a dimuon resonance at 9.5 GeV in 400-GeV proton-nucleus collisions. Phys. Rev. Lett. **39**, 252–255 (1977). [Online]. Available: https://doi.org/10.1103/PhysRevLett.39.252
6. C. Collaboration, Observation of top quark production in $\bar{p}p$ collisions with the collider detector at fermilab. Phys. Rev. Lett. **74**, 2626–2631 (1995). [Online]. Available: https://doi.org/10.1103/PhysRevLett.74.2626
7. D. Collaboration, Observation of the top quark. Phys. Rev. Lett. **74**, 2632–2637 (1995). [Online]. Available: https://doi.org/10.1103/PhysRevLett.74.2632
8. Particle Data Group, Review of particle physics. PTEP **2022**, 083C01 (2022). https://doi.org/10.1093/ptep/ptac097
9. LHC Top Working Group, LHC top working group summary plots. [Online]. Available: https://twiki.cern.ch/twiki/bin/view/LHCPhysics/LHCTopWGSummaryPlots
10. B. Fuks, K. Hagiwara, K. Ma, Y.-J. Zheng, Signatures of toponium formation in LHC run 2 data. Phys. Rev. D **104**, 034023 (2021). [Online]. Available: https://doi.org/10.1103/PhysRevD.104.034023
11. D0 Collaboration, Useful diagrams of top signals and backgrounds. [Online]. Available: https://www-d0.fnal.gov/Run2Physics/top/top_public_web_pages/top_feynman_diagrams.html
12. S. Groote, J. Körner, K. Schilcher, N. Nasrallah, QCD sum rule determination of α(MZ) with minimal data input. Phys. Lett. B **440**(3–4), 375–385 (1998). [Online]. Available: https://doi.org/10.1016/s0370-2693(98)01107-1
13. T. Stelzer, S. Willenbrock, Spin correlation in top-quark production at hadron colliders. Phys. Lett. B **374**(1–3), 169–172 (1996). [Online]. Available: https://doi.org/10.1016/0370-2693(96)00178-5
14. G. Mahlon, S. Parke, Angular correlations in top quark pair production and decay at hadron colliders. Phys. Rev. D **53**, 4886–4896 (1996). [Online]. Available: https://doi.org/10.1103/PhysRevD.53.4886
15. A. Brandenburg, Z. Si, P. Uwer, Qcd-corrected spin analysing power of jets in decays of polarized top quarks. Phys. Lett. B **539**(3), 235–241 (2002). ISSN: 0370-2693. https://doi.org/10.1016/S0370-2693(02)02098-1. [Online]. Available: https://www.sciencedirect.com/science/article/pii/S0370269302020981

16. W. Bernreuther, D. Heisler, Z.-G. Si, A set of top quark spin correlation and polarization observables for the LHC: standard model predictions and new physics contributions. J. High Energy Phys. **2015** (2015). https://doi.org/10.1007/JHEP12(2015)026
17. J. Linacre, Spin correlations in top physics at ATLAS and CMS in run 2 (2019). https://doi.org/10.48550/ARXIV.1905.08634. [Online]. Available: https://arxiv.org/abs/1905.08634
18. CMS Collaboration, Measurement of the top quark polarization and $t\bar{t}$ spin correlations using dilepton final states in proton-proton collisions at $\sqrt{s} = 13$ TeV. Phys. Rev. D **100**(7), 072002 (2019). Preliminary results. https://doi.org/10.1103/PhysRevD.100.072002. arXiv: 1907.03729. [Online]. Available: https://cds.cern.ch/record/2681777
19. ATLAS Collaboration, Measurements of top-quark pair spin correlations in the $e\mu$ channel at $\sqrt{s} = 13$ TeV using pp collisions in the ATLAS detector. Eur. Phys. J. C **80**(8), 754 (2020). Submitted to Eur. Phys. J. https://doi.org/10.1140/epjc/s10052-020-8181-6. arXiv: 1903.07570. [Online]. Available: http://cds.cern.ch/record/2667501

Chapter 4
Monte Carlo Event Simulation

Abstract This chapter provides an overview of the Monte Carlo (MC) simulation techniques used to model proton-proton (pp) collisions. The importance of accurate event simulation for interpreting experimental data is discussed. The stages of simulation are summarized, emphasizing the use of probabilistic and phenomenological models for simulating particle interactions and detector effects.

In particle physics, completely analytical simulations are far too complicated and computationally expensive to be feasible. Instead, numerical Monte Carlo (MC) techniques employ a combination of analytical calculations and both probabilistic and phenomenological models to simulate pp collisions, the production, decay, and showering of particles, as well as the detector response. MC simulations are used to optimize the analysis of data, study the performance of the detector, and compare the results of experiments to theoretical predictions. These simulations are not perfect, as they do contain approximations, modeling assumptions, and systematic uncertainties, but they are sufficiently accurate to make percent precision measurements possible. A schematic representation of the different steps of MC simulation for pp collisions is shown in Fig. 4.1.

To minimize statistical uncertainties, MC simulations are usually performed with statistics much higher than in data recorded by the detector, and can be reweighted to match the luminosity of recorded data:

$$w = \frac{\mathcal{L} \cdot \sigma_{\text{Process}}}{N_{\text{Gen}}} \tag{4.1}$$

where w is the global event weight, $\mathcal{L}$ is the measured integrated luminosity of recorded data, σ_{Process} is the cross-section of the simulated process, and N_{Gen} is weighted sum of simulated events.

J. R. Thieman, *Top Quark Spin Correlations with the CMS Detector*,
Springer Theses, https://doi.org/10.1007/978-3-031-76117-1_4

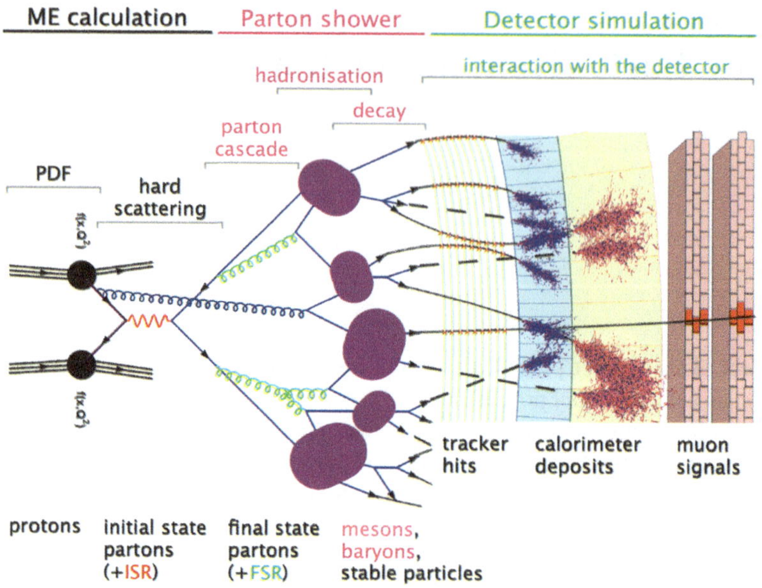

Fig. 4.1 A schematic representation of the different steps of MC simulation for pp collisions [1]

4.1 Parton Distribution Functions

For inelastic pp collisions at the LHC, partons within the proton are at sufficient energies to be asymptotically free, and the hard process can be treated as a direct parton-parton interaction. Protons are composed of not only three valence quarks but also gluons and a sea of virtual quarks. The proton parton distribution functions (PDF), shown in Fig. 4.2, give the probability density f of finding a specific parton within the proton with a certain fraction x of the proton's longitudinal momentum at factorization scale μ_F. The PDFs cannot be calculated from first principles and are determined experimentally by fitting a combination of data from different types of deep inelastic scattering and hadron collision measurements. This measurement uses the NNPDF3.1 NLO set of PDFs [2].

4.2 Matrix Element Calculation and Production Cross-Section

In an event, a hard interaction is a process in which heavy objects are created or a large momentum transfer occurs. For hard parton scattering processes at high energy proton colliders $\alpha_S << 1$, so partonic cross sections can be expanded in a power series expansion of α_S and be calculated in fixed-order perturbation theory.

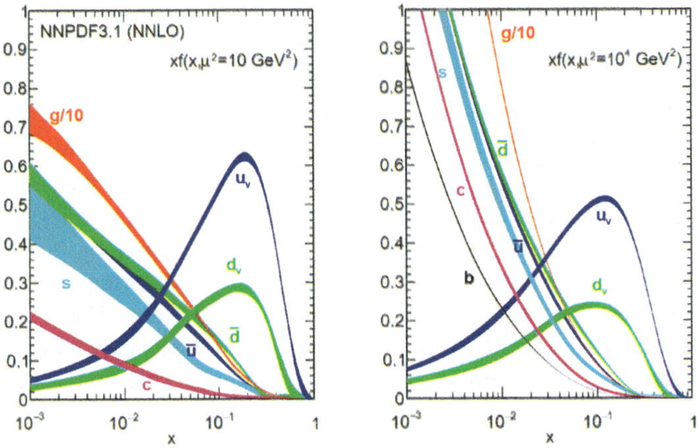

Fig. 4.2 The proton PDF gives the probability density f of finding a specific parton within the proton with a certain fraction x of the proton's longitudinal momentum. Shown for scales $\mu_F = 10\,\text{GeV}^{-2}$ (Left) and at $\mu_F = 100 \times 10^2\,\text{GeV}^{-2}$ (Right), according to the NNPDF collaboration [2]. At high x the PDF is dominated by valence quarks, while at low x it is dominated by gluons and sea quarks

Perturbative calculations to accuracies $\mathcal{O}(\alpha_S^2)$, $\mathcal{O}(\alpha_S^3)$, and $\mathcal{O}(\alpha_S^4)$ are referred to as LO, NLO, and NNLO, respectively. The partonic cross-sections are convoluted with the proton PDFs to obtain pp collision production cross-sections [3]. The cross-section or the scattering process $IJ \to X$ can be expressed as:

$$\sigma_{IJ\to X}(s, \mu_F, \mu_R)$$

$$= \sum_{i,j=g,q,\bar{q}} \int_0^1 \mathrm{d}x_i\, \mathrm{d}x_j\, f_{i/I}\left(x_i, \mu_F^2\right) f_{j/J}\left(x_j, \mu_F^2\right) \cdot \hat{\sigma}_{ij\to X}\left(\hat{s}, \mu_F, \mu_R, \alpha_S\right) \tag{4.2}$$

$$= \sum_{i,j=g,q,\bar{q}} \int_0^1 \mathrm{d}x_i\, \mathrm{d}x_j \int \mathrm{d}\Phi_X\, f_{i/I}\left(x_i, \mu_F^2\right) f_{j/J}\left(x_j, \mu_F^2\right)$$

$$\cdot \frac{1}{2\hat{s}} \left|\mathcal{M}_{ij\to X}\right|^2 (\Phi_X, \mu_F, \mu_R, \alpha_S) \tag{4.3}$$

where $i(j)$ are the parton constituents of proton $I(J)$, $f_{i/I(j/J)}$ are the parton $i(j)$ PDF of proton $I(J)$, $x_{i(j)}$ are the parton $i(j)$ momentum fraction of proton $I(J)$, μ_F is the factorization scale, μ_R is the renormalization scale, $\sigma_{ij\to X}$ are the exclusive partonic cross-sections for X production, $\hat{s} = s \cdot x_i \cdot x_j$ is the squared center-of-mass energy of the parton collision, $\sqrt{s} = 13\,\text{TeV}$ is the center-of-mass energy of the proton-proton collision, α_S is the coupling constant of the strong interaction, Φ_X is the final state phase space, and $\left|\mathcal{M}_{ij\to X}\right|^2$ is the matrix element (ME) squared. The factorization scale μ_F is conventionally set to the energy scale of the high-energy protons in the collision. The renormalization scale μ_R is the arbitrarily chosen high-energy cutoff for the renormalization procedure, used in perturbation theory to separate divergences from higher-order corrections. The ME is calculated by summing over the contributions from the relevant fix-order Feynman diagrams ($\mathcal{M}_{ij\to X} = \sum_k \mathcal{F}^{(k)}_{ij\to X}$), summing over the outgoing spin and color quantum numbers, and averaging over the incoming ones. At LO, ME calculations only include the interaction between incoming partons, but higher-order ME calculations will additionally include real emission and virtual loop corrections. The real emission corrections correspond to radiated partons, which are classified as initial state radiation (ISR) and final state radiation (FSR) depending on whether they are emitted by the incoming or outgoing particles. These measurements uses the Positive Weight Hard Event Generator v2 (POWHEGV2) event generator, which performs ME calculations to NLO accuracy and accounts for polarizations in $t\bar{t}$ production and top quark decays.

4.3 Parton Shower

Parton shower (PS) models account for higher-order QCD and QED radiative emissions beyond the accuracy of fixed-order perturbative approximations. Showers of outgoing partons are created as partons radiate gluons and gluons radiate other gluons or produce $q\bar{q}$ pairs. In this measurement, the PS is modeled using PYTHIA8. In PYTHIA8 the radiative emissions are modeled as a Markov chain, where each emission is treated as independent, with the probability of an additional emission being determined by the splitting functions and the kinematics of the partons involved [4]. The algorithm continues to iteratively split the partons in a cascade of emissions until they reach an energy scale where perturbation theory breaks down.

A matching and merging procedure is used to avoid double-counting radiative emissions from the ME calculation and the PS models [5]. Two general NLO matching methods exist, namely POWHEG and MC@NLO, available in POWHEGV2 and automated in MADGRAPH5_AMC@NLO respectively [4]. In POWHEGV2, the h_{damp} parameter controls the matching procedure by modifying the scale at which radiative emissions from the PS are discarded in favor of the ME calculations. Lower (higher) values of h_{damp} result in softer (harder) radiated emissions and less (more) events with high p_{T} jets. MLM and UMEPS are tree-level methods for merging, while FXFX, which combines MC@NLO matching with MLM matching, and UNLOPS, which is an extension of UMEPS, are for NLO merging [4]. UNLOPS is the preferred NLO merging scheme in PYTHIA8, but the FXFX procedure is used for NLO MC simulations when the ME calculations were performed by MADGRAPH5_AMC@NLO.

For pp collisions, besides the parton interaction in the hard process, the remaining constituent partons of the colliding protons, referred to as the beam-

beam remnants (BBR), can also interact. These multiple parton interactions (MPI) can produce additional independent scattering processes that contribute to the underlying event (UE). PS models in event generators have adjustable parameters to control the behavior of event modeling, and a set of these parameters, which has been adjusted to better fit some aspects of the data, is referred to as a UE tune [6]. The CP5 UE tune was chosen for all simulated samples for this measurement.

4.4 Hadronization

Hadronization is the direct consequence of color confinement and the running of α_S at low energy scales where QCD is non-perturbative. As non-perturbative QCD is unsolved, hadronization is simulated using phenomenological models that describe the process of forming colorless hadrons from quarks and gluons. In PYTHIA8, it is based on the Lund string model, which assumes that the potential energy of color charge dipole fields grows linearly ($V(r) = \kappa r$ with $\kappa \approx 1\,\mathrm{GeV\,f^{-1}\,m}$) with the separation of the charges [7]. As color charges (q and $\bar{q}'$) separate, the potential energy increases beyond the threshold that the color flux tube snaps, and a new color charge pair (q' and $\bar{q}'$) emerge. After the snap, the system consists of two separate color charge dipole fields ($q\bar{q}''$ and $q''\bar{q}'$). Further snaps continue until the potential energy of the tubes is not sufficient to produce new color charge pairs.

In these models, individual partons do not hadronize independently, but rather color-connected systems of partons hadronize collectively [3]. The scenario in which color fields connecting different partons rearrange themselves after the collision and before the formation of final hadrons is referred to as color reconnection (CR). To more accurately describe the data, phenomenological CR models are used to simulate color flow. PYTHIA8 uses an MPI-based CR scheme in which there is a probability that the partons of the lower-p_T proton are added to the strings defined by the higher-p_T proton in such a way as to give the smallest total string length.

4.5 Detector Response

The interactions of the MC-generated final state particles with the CMS detector are simulated with the GEometry ANd Tracking (GEANT4 v10.4.3) toolkit. GEANT4 divides the trajectory of particles traversing the detector into small, finite steps and calculates the probability that the particles will interact with the material in the detector based on their properties and the properties of the material they are passing through [8]. A complete set of physics models for the electromagnetic, strong, and weak interactions are included in GEANT4. If an interaction occurs, GEANT4 calculates the resulting energy deposition, generates any secondary particles that are produced, and simulates the response of the readout electronics and trigger system including the effects of noise and other sources of background. To provide an accurate simulation, GEANT4 uses a detailed description of the detector geometry and materials, calibrations, and electronics performance including dead channels. Experimental data is used to parameterize detector response and validate the simulation [9]. The collection of GEANT4-simulated detector responses for an event is stored in the raw CMS event data model (EDM) format and can be reconstructed exactly as actual events recorded by the CMS detector.

References

1. N. Bartosik, Associated top-quark-pair and b-jet production in the dilepton channel at $\sqrt{s} = 8$ TeV as test of QCD and background to tt+Higgs production. Universität Hamburg, Diss. (2015). Dr. Universität Hamburg, Hamburg (2015), p. 312. https://doi.org/10.3204/DESY-THESIS-2015-035. [Online] Available: https://bib-pubdb1.desy.de/record/222384
2. R.D. Ball, V. Bertone, S. Carrazza, et al., Parton distributions from high-precision collider data. Parton distributions from high-precision collider data. Eur. Phys. J. C **77**(10), 663 (2017). https://doi.org/10.1140/epjc/s10052-017-5199-5. arXiv: 1706.00428. [Online]. Available: https://cds.cern.ch/record/2267455
3. A. Buckley, J. Butterworth, S. Gieseke, et al., General-purpose event generators for LHC physics. Phys. Rep. **504**(5), 145–233 (2011). ISSN: 0370-1573. https://doi.org/10.1016/j.physrep.2011.03.005. [Online]. Available: https://www.sciencedirect.com/science/article/pii/S0370157311000846
4. C. Bierlich, S. Chakraborty, N. Desai, et al., A comprehensive guide to the physics and usage of pythia 8.3 (2022). https://doi.org/10.48550/ARXIV.2203.11601. [Online]. Available: https://arxiv.org/abs/2203.11601
5. S. Frixione, P. Nason, C. Oleari, Matching nlo qcd computations with parton shower simulations: the powheg method. J. High Energy Phys. **2007**(11), 070 (2007). [Online]. Available: https://doi.org/10.1088/1126-6708/2007/11/070
6. CMS Collaboration, Extraction and validation of a new set of CMS PYTHIA8 tunes from underlying-event measurements. Eur. Phys. J. C **80**(1), 4 (2020). All the figures and tables can be found at http://cms-results.web.cern.ch/cms-results/public-results/publications/GEN-17-001 (CMS Public Pages). https://doi.org/10.1140/epjc/s10052-019-7499-4. arXiv: 1903.12179. [Online]. Available: https://cds.cern.ch/record/2669320
7. T. Sjöstrand, S. Ask, J.R. Christiansen, et al., An introduction to Pythia 8.2. Comput. Phys. Commun. **191**, 159–177 (2015). ISSN: 0010-4655. https://doi.org/10.1016/j.cpc.2015.01.024. [Online]. Available: https://www.sciencedirect.com/science/article/pii/S0010465515000442
8. S. Agostinelli, J. Allison, K. Amako, et al., Geant4a simulation toolkit. Nucl. Instrum. Methods Phys. Res. A **506**(3), 250–303 (2003). ISSN: 0168-9002. https://doi.org/10.1016/S0168-9002(03)01368-8. [Online]. Available: https://www.sciencedirect.com/science/article/pii/S0168900203013688
9. CMS Collaboration, CMS Physics: Technical Design Report Volume 1: Detector Performance and Software (Technical design report. CMS). Geneva: CERN, 2006, There is an error on cover due to a technical problem for some items. [Online]. Available: https://cds.cern.ch/record/922757

Chapter 5
Datasets, Event Selection, and $t\bar{t}$ Kinematic Reconstruction

Abstract This chapter details the object reconstruction, event selection criteria, CMS detector data sets used for measurement, and event yields associated with the measurement results of this dissertation. Methods for identifying and measuring the properties of particles such as electrons, muons, jets, and missing transverse energy are explained. Kinematic reconstruction of top quark pairs and the strategies for distinguishing signal events from background noise are also described.

The signal process for this measurement is the production of top quark pairs followed by top quark decays $t \to W^+b$ and $\bar{t} \to W^-\bar{b}$, and subsequent leptonic W boson decays into final state muons $W \to \mu\nu$ and electrons $W \to e\nu$, both "prompt" and "via tau," i.e. via the decay of the W boson into a tau $W \to \tau\nu$ and its subsequent leptonic decay into a final state electron $\tau \to e\nu$ or muon $\tau \to \mu\nu$. The analysis has been carried out using the CMS EDM and official software framework for event generation, simulation, and reconstruction, following recommendations for ultra-legacy Run II analyses by the Physics Performance and Datasets (PPD) and the Physics Data And Monte Carlo Validation (PdmV) CMS groups.

5.1 Datasets

5.1.1 Recorded Datasets

These measurements are performed using $19.50\,\text{fb}^{-1} \pm 1.2\%$ [1], $16.81\,\text{fb}^{-1} \pm 1.2\%$ [1], $41.48\,\text{fb}^{-1} \pm 2.3\%$ [2], and $59.83\,\text{fb}^{-1} \pm 2.5\%$ [3] (Total: $137.7\,\text{fb}^{-1} \pm 1.6\%$) of data collected with the CMS experiment at the LHC with $\sqrt{s} = 13\,\text{TeV}$, during 2016, 2017, and 2018 respectively. Only the runs and luminosity sections that had good functioning of every CMS sub-detector were selected for analysis. The measurements are performed using ultra legacy CMS data sets, which are comprehensively documented, designed to be useful for a wide range of studies, preserved for long-term access, and include high-level physics objects such as

J. R. Thieman, *Top Quark Spin Correlations with the CMS Detector*,
Springer Theses, https://doi.org/10.1007/978-3-031-76117-1_5

reconstructed leptons, jets, and missing transverse energy. The 2016 data set is separated into "preVFP" and "postVFP", partitioning the year into the periods before and after heavily ionizing particle mitigation (HIPM) was implemented, while no such separation for the 2017 or 2018 data sets exist.

5.1.2 *Simulated Datasets*

MC simulations (described in Chap. 4) are used to estimate signal and background contributions. The dileptonic $t\bar{t}$ signal sample is produced using the POWHEGV2 event generator with NLO ME calculations. PS and hadronization of the $t\bar{t}$ signal sample is performed using PYTHIA8. The matrix-element jets are matched to parton showers using the POWHEGV2 method. The MC simulations assume that $m_t = 172.5$ GeV, and the PDFs are described using NNPDF3.1.

The major sources of background contributions are semi-leptonic $t\bar{t}$, fully hadronic $t\bar{t}$, dileptonic decays of single top quarks in association with a W boson (tW and $\bar{t}W$), $t\bar{t}$ in association with W/Z bosons, Z+jets, W+jets, and WW, WZ, ZZ diboson processes. A summary of the MC simulations for signal and background processes is shown in Table 5.1, including the ME generator, PS algorithm, and cross-section used for luminosity scaling.

5.2 Particle-Flow Reconstruction and Event Selection

The CMS detector consists of several sub-detector layers that exploit the different properties of particles to identify them and measure their energy-momenta and trajectories. For reconstructing particle objects from their interaction with the detector, the CMS experiment has developed a holistic particle-flow (PF) reconstruction algorithm that combines the measurements from all relevant sub-detector layers [4]. The algorithm links the charged-particle tracks measured in the inner silicon tracker with the energy deposits in the calorimeters and the muon tracks measured in the outer muon tracking system. As there are trade-offs between reconstruction efficiency and misidentification rate, the criteria for classifying reconstructed objects are determined by dedicated CMS Physics Object Groups (POG) which perform expert analyses to optimize the efficiency and purity of object identification.

The main particles that are directly detected by CMS are muons, electrons, photons, charged hadrons (e.g. protons, pions, kaons), and neutral hadrons (e.g. neutrons). An illustration of the signatures left behind by these particles in the various sub-detectors is shown in Fig. 5.1. The PF algorithm identification and reconstruction sequence first identifies and reconstructs muon candidates, then electron and photon candidates, and finally charged and neutral hadron candidates. After each step, tracks and clusters of reconstructed objects are removed from further consideration for subsequent steps.

Table 5.1 A summary of the MC simulations for signal and background processes including the ME generator, PS algorithm, and cross-section used for luminosity scaling

Process	ME (Matching)	PS	σ [pb]
$t\bar{t}$ (Dileptonic)	POWHEGV2	PYTHIA8	831.76×0.10706
$t\bar{t}$ (Semi-Leptonic)	POWHEGV2	PYTHIA8	831.76×0.44113
$t\bar{t}$ (Hadronic)	POWHEGV2	PYTHIA8	831.76×0.45441
$t\bar{t} + W$ (Leptonic W)	MADGRAPH5_AMC@NLO+MADSPIN	PYTHIA8	0.2043
$t\bar{t} + W$ (Hadronic W)	MADGRAPH5_AMC@NLO+MADSPIN	PYTHIA8	0.4062
$t\bar{t} + Z$ (Leptonic Z)	MADGRAPH5(MLM)	PYTHIA8	0.2529
$t\bar{t} + Z$ (Hadronic Z)	MADGRAPH5(MLM)	PYTHIA8	0.5297
$t + W$ (Dileptonic)	POWHEGV2	PYTHIA8	35.85×0.54559
$\bar{t} + W$ (Dileptonic)	POWHEGV2	PYTHIA8	35.85×0.54559
W + Jets	MADGRAPH5(MLM)	PYTHIA8	61526.7
Z/γ^* + Jets ($10\,\text{GeV} < m_{\ell\bar{\ell}} < 50\,\text{GeV}$)	MADGRAPH5(MLM)	PYTHIA8	22635.1
Z/γ^*+Jets ($m_{\ell\bar{\ell}} > 50\,\text{GeV}$)	MADGRAPH5(MLM)	PYTHIA8	0.5×6225.4
Z/γ^*+Jets ($m_{\ell\bar{\ell}} > 50\,\text{GeV}$)	MADGRAPH5_AMC@NLO(FXFX)	PYTHIA8	0.5×6225.4
WW	PYTHIA8	PYTHIA8	118.7
WZ	PYTHIA8	PYTHIA8	47.13
ZZ	PYTHIA8	PYTHIA8	16.523

For the measurements presented in this dissertation, the reconstruction of the different objects is performed using the PF algorithm and the object identification criteria follow the recommendations for ultra-legacy Run II analyses by the CMS Top Physics Analysis Group (Top PAG). The $t\bar{t}$ dilepton final state is characterized by the presence of at least two high-p_{T} isolated leptons with opposite electric charge, large MET ($E_{\mathrm{T}}^{\mathrm{miss}}$), and two jets created from the hadronization of b-quarks.

5.2.1 Triggers

To maximize the trigger efficiency, dilepton data streams and single-lepton streams are both used for this measurement. In order to ensure no double-counting of events, events passing the dilepton triggers are vetoed when processing the single lepton data streams. Approximately 10% of dilepton events that failed to pass the dilepton trigger requirements are recovered by including the single lepton data streams. The trigger efficiency is measured in data as a function of the lepton p_{T} and used to correct the MC simulations.

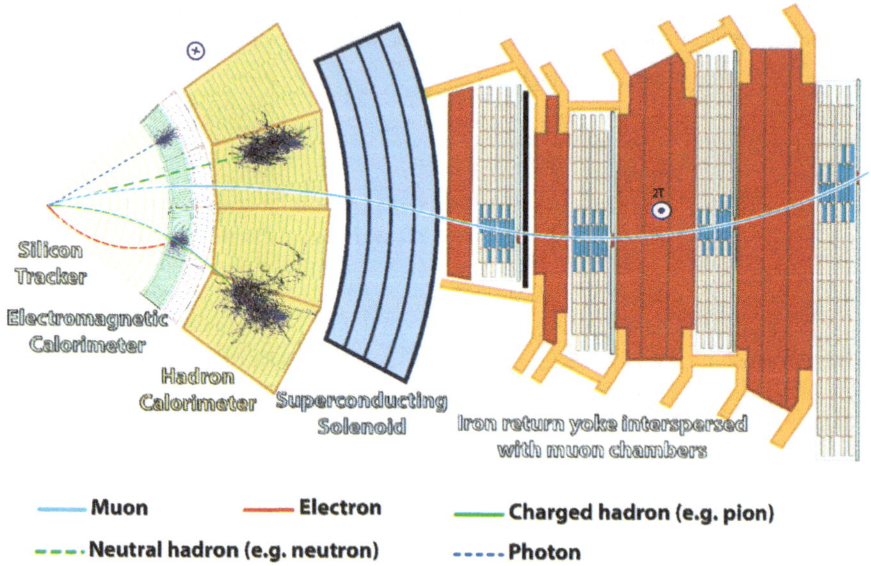

Fig. 5.1 Muons, electrons, photons, charged hadrons, and neutral hadrons are directly detected by CMS and are identified by the signatures they leave behind in the various sub-detectors [4]

5.2.2 Track and Vertex Reconstruction, Primary Vertex Requirements, and Pileup Corrections

Every 25 ns, an average bunch crossing results in 1000 or more particles from 20 or more *pp* collisions [5]. Track reconstruction is the determination of final state particle trajectories using the ionization trails left behind in the tracker sub-detector layers and is crucial for the identification and momenta measurement of charged particles. The situation where multiple *pp* collisions (vertices) occur within the same bunch crossing is referred to as PU, and vertexing is the determination of vertices from the intersection of trajectories. The primary vertex (PV) of an event is the vertex corresponding to the hard-scattering interaction in the collision and is identified as the vertex with the highest sum of squared transverse momenta of its associated tracks. Vertexing is also important for the identification of secondary vertices (SV), i.e. vertices separated from the beamline by $\gtrsim$1 mm, which are usually attributed to the decay of short-lived heavy flavor mesons.

In the PF algorithm, iterative Kalman filtering (KF) is used to reconstruct the tracks and vertices of charged particles [4]. KF assumes that the residuals between the predicted and measured trajectories are Gaussian distributions. The algorithm is seeded with tracks generated from two hits in consecutive layers in the pixel detector. The KF fits the expected trajectory of the particles based on measured charge and momentum, accounts for multiple scattering and energy loss of particles in the detector material, and updates the track parameters based on measurements

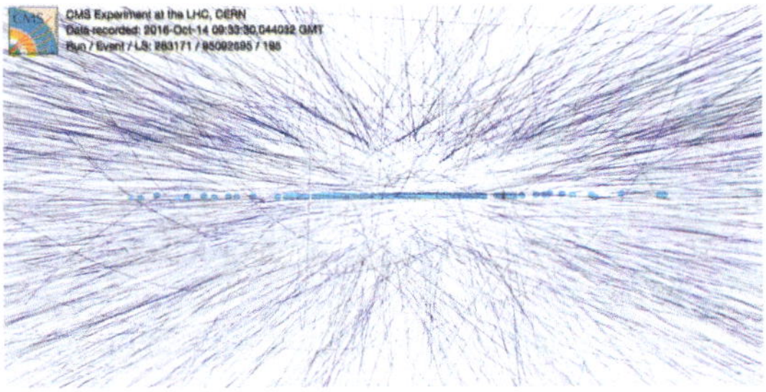

Fig. 5.2 A visualization of the reconstructed tracks and identified vertices for an event with high PU recorded by the CMS detector in 2016 [6]

in each detector layer. A visualization of the reconstructed tracks and identified vertices for an event with high PU recorded by the CMS detector in 2016 is shown in Fig. 5.2.

For the measurements presented in this dissertation, the PV of an event is required to be associated with at least four tracks and be in the vicinity of the nominal interaction point with $|r| < 2\,\text{cm}$ and $|z| < 24\,\text{cm}$. Charged-hadron subtraction (CHS) is used to remove charged PU contributions and the L1FastJet algorithm is applied to subtract the remaining neutral contributions [7]. The number of PU events in MC simulations is typically based on an estimate of the expected number of interactions per bunch crossing and PU re-weighting is performed using the instantaneous luminosity per bunch crossing in data, and the total *pp* inelastic cross section of 69.2 mb, to correct the PU distribution of the MC simulation.

5.2.3 *Muons*

Muons leave tracks in the silicon tracker, their trajectories are curved due to the presence of the magnetic field of the solenoid, and they penetrate through the calorimeters and solenoid to leave ionization trails in the chambers of the muon system. As is implied by its name, the CMS experiment was designed to detect muons with high reconstruction efficiency and low misidentification rate. Muons that are reconstructed independently in both the inner tracking detector and the outer muon system are referred to as global muons. A global fit of the muon momentum using both sub-detectors has momentum resolution of 1–3% for low momentum muons ($20\,\text{GeV} < p_T < 200\,\text{GeV}$), depending on η, and 5% for high momentum muons ($\sim$1 TeV) [5]. If hadron shower remnants "punch-through" the HCAL and solenoid into the outer muon system, they may be misreconstructed as muons, but additional criteria can be used to balance the efficiency and purity of muon identification.

For the measurements presented in this dissertation, PF muon candidates are required to have a transverse momentum $p_T > 20\,\text{GeV}$ and a pseudorapidity restricted to the coverage of the inner tracker $|\eta| < 2.4$. They are also required to be reconstructed as global muons and fulfill tight selection criteria chosen to ensure high purity, provide good p_T measurements, and suppress punch-through fake-muons and muons produced from hadron decays. To remove leptons overlapping with jets, PF muon candidates are required to fulfill the isolation condition $I_{Rel}^{PF} < 0.15$. Corrections are also applied that scale the raw energy measurements and smear the muon resolutions to match the accuracy and precision of reconstructed muons in MC simulation to those in recorded data. Identification and isolation efficiency corrections for muons are also applied and are measured as a function of p_T and η using a "tag-and-probe" method with an orthogonal dataset.

5.2.4 *Electrons*

Electrons leave ionization trails in the silicon tracker, their trajectories are curved due to the presence of the magnetic field of the solenoid, and their energy is deposited in the ECAL. Electron tracks are seeded from the ECAL clusters and extrapolated back to the inner silicon tracker. Gaussian Sum Filtering (GSF), which assumes that the residuals between the predicted and measured trajectories are sums of multiple Gaussian distributions, is used to reconstruct electrons. Due to the total radiation length of the inner silicon tracker, the probability that an electron emits a bremsstrahlung photon by interacting with tracker material is about 85% [4]. Thus the ECAL clusters from the electron and the bremsstrahlung photons in a window around the electron direction are grouped into a "supercluster." For non-isolated electrons, such as those produced in jets, the superclustering can result in large inefficiencies, so the ECAL seeding criteria are restricted by isolation requirements.

For the measurements presented in this dissertation, PF electron candidates are required to have transverse momentum $p_T > 20\,\text{GeV}$ and a pseudorapidity restricted to the coverage of the inner tracker $|\eta| < 2.4$. The gap between the barrel and endcap region of the ECAL ($1.4442 < |\eta_{\text{sc}}| < 1.5660$) is excluded, where η_{sc} is the pseudorapidity of the ECAL supercluster. To ensure high purity, electron selection uses tight identification and isolation criteria. The electron isolation considers photons, neutral hadrons, and charged hadrons as identified by the PF algorithm in a η-ϕ space cone of $\Delta R < 0.3$ around the electron. Corrections are also applied that scale the raw energy measurements and smear the electron resolutions to match the accuracy and precision of reconstructed electrons in MC simulation to those in recorded data. Identification and reconstruction efficiency corrections for electrons are also applied and are measured as a function of p_T and η using a "tag-and-probe" method with an orthogonal dataset.

5.2.5 *Lepton Pair*

Events with a dilepton system consisting of exactly two oppositely-charged leptons passing the electron and muon object selection criteria are accepted for further consideration. If more than two leptons are reconstructed in the event, then the event is vetoed. The leading selected lepton is required to have $p_{\mathrm{T}} > 25\,\mathrm{GeV}$. The event is then unambiguously classified as e^+e^-, $e^\pm\mu^\mp$, or $\mu^+\mu^-$ depending on the type of the selected lepton pair. The invariant mass of the selected lepton pair is required to be larger than 20 GeV to suppress background events from decays of heavy-flavor resonances and Drell-Yan processes. Moreover, in the e^+e^- and $\mu^+\mu^-$ decay channels, events are rejected if the dilepton invariant mass is within the vicinity of the Z boson mass $76\,\mathrm{GeV} < m_{\ell\bar{\ell}} < 106\,\mathrm{GeV}$, where background from Z boson production is dominant.

Control plots of the dilepton invariant mass and jet multiplicity distributions for combined 2016preVFP, 2016postVFP, 2017, 2018 eras and combined e^+e^-, $e^\pm\mu^\mp$, $\mu^+\mu^-$ channels after selection of two isolated leptons passing the trigger requirements and $m_{\ell\bar{\ell}} > 20\,\mathrm{GeV}$ are shown in Fig. 5.5. For the jet multiplicity distribution, events in the e^+e^- and $\mu^+\mu^-$ channels with $76\,\mathrm{GeV} < m_{\ell\bar{\ell}} < 106\,\mathrm{GeV}$ have also been rejected.

5.2.6 Jets

Quarks and gluons produced in pp collisions hadronize into collimated showers of particles called jets. Jets primarily consist of hadrons, but also can contain photons and leptons when hadrons in the jet radiate or decay before reaching the detector. Charged hadrons leave tracks in the silicon tracker, have curved trajectories due to the presence of the magnetic field of the solenoid, deposit some energy in the ECAL, and exhaust their energy in the HCAL, while neutral hadrons do not leave tracks in the silicon tracker and exhaust their energy in the HCAL. The granularity of the inner silicon tracker is sufficiently fine to provide separation of individual particle trajectories within jets, and energy clusters are deposited in the calorimeters from the particles in the shower.

The anti-k_T jet clustering algorithm groups reconstructed PF objects into jets by merging the pair of particles with the smallest distance, calculating new distances between the merged object and the remaining particles, and repeating this process until all particles have been merged into jet objects [8]. For anti-k_T jet clustering, the distance d_{ij} between two particles i and j is defined as:

$$d_{\mathrm{ij}} = min(1/p_{T,i}^2, 1/p_{T,j}^2) \times \Delta R_{ij}{}^2/R^2 \tag{5.1}$$

where ΔR is the angular separation between objects, defined in Eq. 2.7, and R is the radius parameter that determines the size of individual jets. The PF algorithm reduces the amount of energy that is mistakenly assigned to a jet due to PU by removing clusters attributed to tracks originating at PU vertices. Due to complications such as assumptions about particle separation in the clustering algorithm, energy loss in the detector material, detector noise, PU, detector granularity, the association of neutral hadrons with interaction vertices, and the stochastic nature of particle showers in the HCAL, reconstructed jet energies and trajectories are typically associated with large systematic uncertainties.

For the measurements presented in this dissertation, jets are clustered from reconstructed PF candidates using the anti-k_T clustering algorithm with radius parameter $R = 0.4$ (AK4). Events are required to have at least two jets with transverse momentum $p_{\mathrm{T}} > 30\,\mathrm{GeV}$ and within the coverage of the inner tracker $|\eta| < 2.4$. Identification criteria is applied to efficiently identify jets (efficiency $\sim$ 98–99%) while rejecting jets with significant lepton fractions (purity $\sim$ 98%). To exclude jets overlapping with selected leptons used in the analysis, a cleaning of leptons from jets is applied if $\Delta R(jet, lepton) < 0.4$. Events with jets in regions of the calorimeter that produced anomalously high or low jet rates are vetoed in MC and recorded data.

Jet energy corrections (JEC) are applied that adjust the jet energy scale (JES) to match the accuracy of reconstructed jet energies in MC simulation to those in recorded data. To reduce the contribution coming from PU, the CHS algorithm removes charged particles and the L1FastJet algorithm removes the remaining neutral contributions from PU vertices before clustering jets [7]. L2L3 MC-truth corrections are applied to both data and MC simulation to correct for the variation of detector resolution as a function of jet η and p_{T}. Any remaining discrepancies in the accuracy of jet energy measurements due to detector response and other effects are eliminated with the application of L2L3Residual corrections. Jet energy resolution (JER) corrections are applied by smearing the jets in MC simulation to match the precision of jets in recorded data.

5.2.7 *Missing Transverse Energy*

For collisions at the LHC, there is no net momentum or force in the transverse direction, so p_{T} is conserved and sums to zero for both incoming protons and outgoing particles after the collision. This property is useful for inferring the production of particles invisible to the detector, i.e. neutrinos, from the missing transverse energy (MET):

$$\vec{E}_T^{miss} = -\sum_i \vec{p}_{T,i} \tag{5.2}$$

the negative vector sum of the transverse momenta for all PF reconstructed particles in an event, illustrated in Fig. 5.3.

For the measurements presented in this dissertation, the calculation of the $E_{\mathrm{T}}^{\mathrm{miss}}$ is based on PF objects, where pileup-per-particle-identification (PUPPI) [7] is used for PU mitigation. PUPPI is an alternative to the PF CHS algorithm which gives weights to particles based on the probability that they come from PU or the PV. The JEC (JES and JER) and lepton energy scale corrections are propagated to the $E_{\mathrm{T}}^{\mathrm{miss}}$.

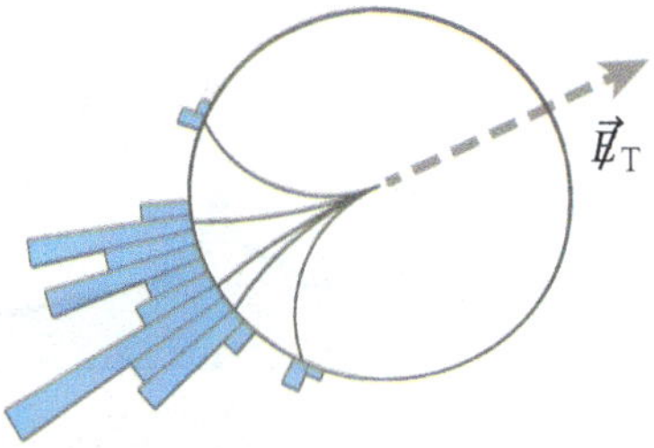

Fig. 5.3 Schematic diagram for $E_{\mathrm{T}}^{\mathrm{miss}}$, the negative vector sum of the transverse momenta for all reconstructed particles in an event [9]

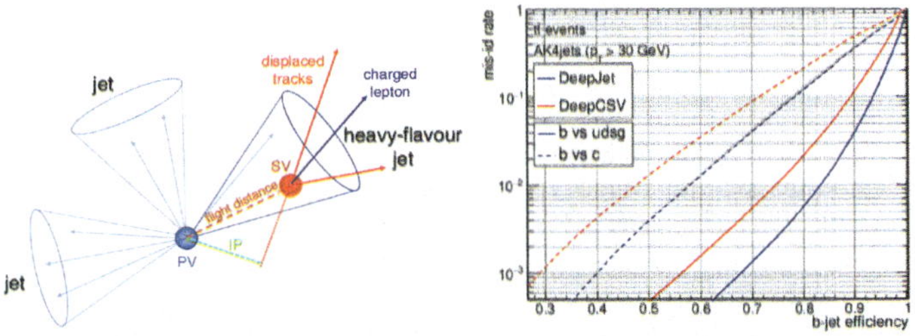

Fig. 5.4 Illustration of a heavy-flavor jet with a SV from the decay of a b or c hadron resulting in charged-particle tracks (including possibly a soft lepton) that are displaced with respect to the PV, and hence with a large impact parameter (IP) value (Left) [11]. Misidentification probability for c and l-jets versus b jet identification efficiency for DeepJet and DeepCSV b tagging algorithms applied to jets in $t\bar{t}$ events (Right) [11]

Events in the e^+e^- and $\mu^+\mu^-$ channels are required to have $E_{\mathrm{T}}^{\mathrm{miss}} > 40\,\mathrm{GeV}$, but no requirement on $E_{\mathrm{T}}^{\mathrm{miss}}$ is applied in the $e^{\pm}\mu^{\mp}$ channel.

Controls plots of the $E_{\mathrm{T}}^{\mathrm{miss}}$ and b-tagged jet multiplicity distributions for combined 2016preVFP, 2016postVFP, 2017, 2018 eras and combined e^+e^-, $e^{\pm}\mu^{\mp}$, $\mu^+\mu^-$ channels after selection of two isolated leptons passing the trigger requirements, $m_{\ell\bar{\ell}} > 20\,\mathrm{GeV}$, $N_{jets} \geq 2$, and rejecting events in the e^+e^- and $\mu^+\mu^-$ channels with $76\,\mathrm{GeV} < m_{\ell\bar{\ell}} < 106\,\mathrm{GeV}$ are shown in Fig. 5.6. For the b-tagged jet multiplicity distribution, events in the e^+e^- and $\mu^+\mu^-$ channels with $E_{\mathrm{T}}^{\mathrm{miss}} < 40\,\mathrm{GeV}$ have also been rejected.

5.2.8 *Tagging of b-Jets*

Selected events are required to have at least one jet tagged as having originated from a b-quark. In this analysis, the DeepJet b-tagging algorithm [10], which uses approximately 650 input variables, divided into four categories (global variables, charged PF candidate features, neutral PF candidate features, and SV features associated with the jet) as inputs for a deep neural network, is used to tag all reconstructed jets in the event. A DeepJet medium working point is used with light (l)-jet mistag efficiency of $\sim$1% (see Fig. 5.4).

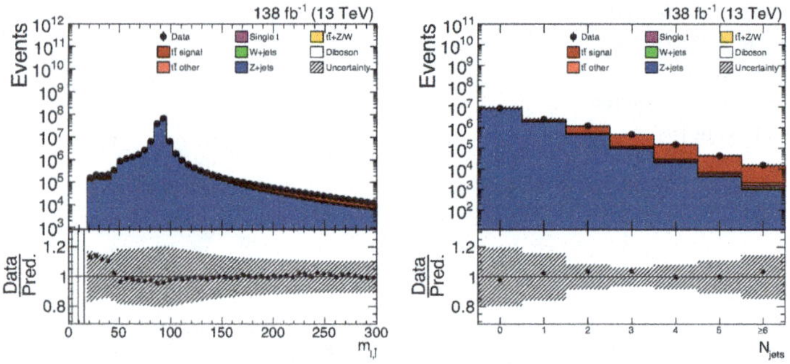

Fig. 5.5 Dilepton invariant mass and jet multiplicity distributions for combined 2016preVFP, 2016postVFP, 2017, 2018 eras and combined e^+e^-, $e^\pm\mu^\mp$, $\mu^+\mu^-$ channels after selection of two isolated leptons passing the trigger requirements and $m_{\ell\bar{\ell}} > 20\,\text{GeV}$. For the jet multiplicity distribution, events in the e^+e^- and $\mu^+\mu^-$ channels with $76\,\text{GeV} < m_{\ell\bar{\ell}} < 106\,\text{GeV}$ have also been rejected. The simulated samples are re-weighted to an integrated luminosity of 137.7fb^{-1}. Trigger efficiency and lepton selection scale factors discussed in Sect. 6.7.1 are applied. No Z+jets scaling has been applied yet. Statistical uncertainties are shown as error bars on the data points, and total systematic uncertainties are indicated by the hatched uncertainty band

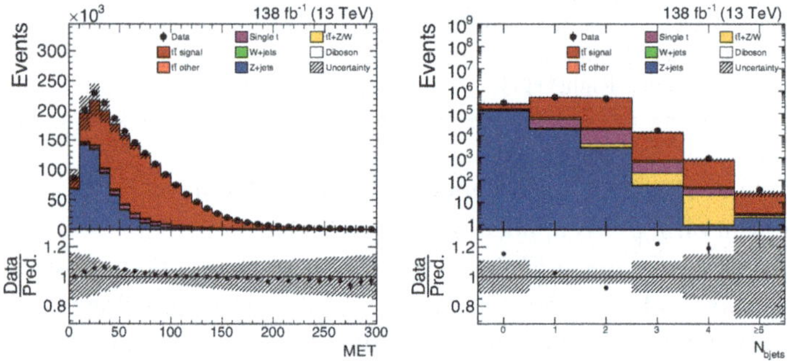

Fig. 5.6 $E_{\text{T}}^{\text{miss}}$ and b-tagged jet multiplicity distributions for combined 2016preVFP, 2016postVFP, 2017, 2018 eras and combined e^+e^-, $e^\pm\mu^\mp$, $\mu^+\mu^-$ channels after selection of two isolated leptons passing the trigger requirements, $m_{\ell\bar{\ell}} > 20\,\text{GeV}$, $N_{jets} \geq 2$, and rejecting events in the e^+e^- and $\mu^+\mu^-$ channels with $76\,\text{GeV} < m_{\ell\bar{\ell}} < 106\,\text{GeV}$. For the b-tagged jet multiplicity distribution, events in the e^+e^- and $\mu^+\mu^-$ channels with $E_{\text{T}}^{\text{miss}} < 40\,\text{GeV}$ have also been rejected. The simulated samples are re-weighted to an integrated luminosity of 137.7fb^{-1}. Trigger efficiency, lepton selection, and b-tagging efficiency scale factors discussed in Sect. 6.7.1 are applied. The Z+jets contributions in MC have been scaled to better describe the number of events observed in recorded data, as discussed in Sect. 5.4. Statistical uncertainties are shown as error bars on the data points, and total systematic uncertainties are indicated by the hatched uncertainty band

5.2.9 Event Selection Control Plots

See Figs. 5.5, 5.6, and 5.7.

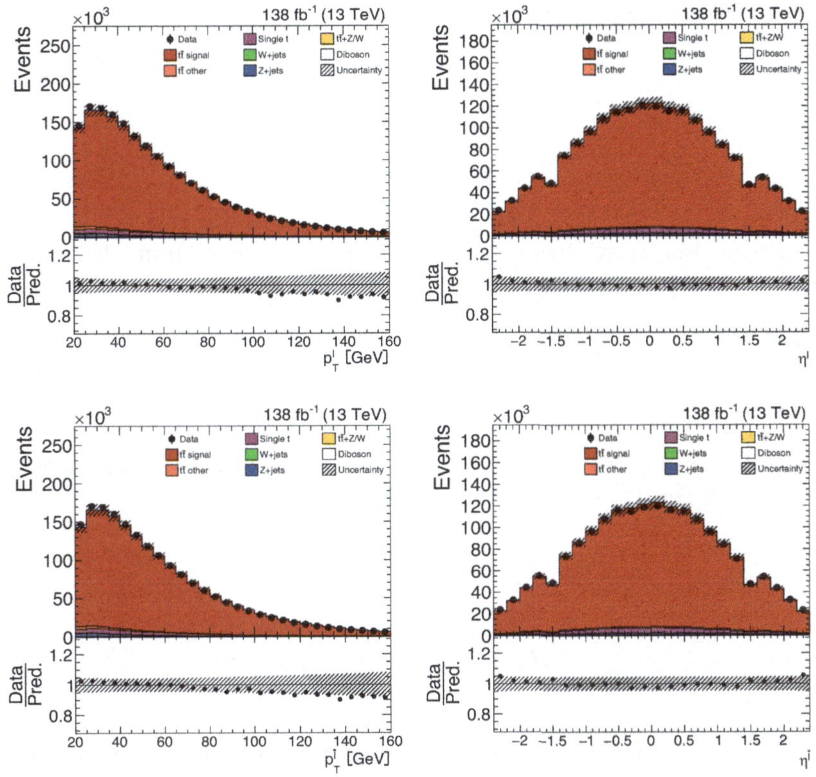

Fig. 5.7 p_T and η distributions for selected leptons and anti-leptons for combined 2016preVFP, 2016postVFP, 2017, 2018 eras and combined e^+e^-, $e^{\pm}\mu^{\mp}$, $\mu^+\mu^-$ channels after all object and event selection criteria have been applied. The simulated samples are re-weighted to an integrated luminosity of 137.7fb^{-1}. Trigger efficiency, lepton selection, b-tagging efficiency, and kinematic reconstruction efficiency scale factors discussed in Sect. 6.7.1 are applied. The Z+jets contributions in MC have been scaled to better describe the number of events observed in recorded data, as discussed in Sect. 5.4. Statistical uncertainties are shown as error bars on the data points, and total systematic uncertainties are indicated by the hatched uncertainty band

5.3 t$\bar{\text{t}}$ Kinematic Reconstruction

5.3.1 Full $t\bar{t}$ Kinematic Reconstruction

To fully pass event selection, an event is also required to have at least one solution to the $t\bar{t}$ kinematic reconstruction. The top quark and anti-quark are fully reconstructed using an analytical method [12], in which eight kinematic constraints are applied to determine the unknown four-momenta of the undetected neutrinos. The constraint equations are constructed from the four-momenta of the objects passing event selection: two leptons, two jets, and the missing transverse energy.

The two neutrinos are assumed to be massless:

$$m_\nu^2 = E_\nu^2 - \sum_{i=x,y,z} p_{\nu_i}^2 = 0 \tag{5.3}$$

$$m_{\bar{\nu}}^2 = E_{\bar{\nu}}^2 - \sum_{i=x,y,z} p_{\bar{\nu}_i}^2 = 0 \tag{5.4}$$

The missing transverse energy is assumed to be entirely from the transverse momenta of the two neutrinos:

$$E_x^{\text{miss}} = p_{\nu_x} + p_{\bar{\nu}_x} \tag{5.5}$$

$$E_y^{\text{miss}} = p_{\nu_y} + p_{\bar{\nu}_y} \tag{5.6}$$

The lepton (anti-lepton) and the anti-neutrino (neutrino) are assumed to have an invariant mass equal to the W boson mass of $m_{W^\pm} = 80.4\,\text{GeV}$:

$$m_{W^+}^2 = \left(E_{\bar{\ell}} + E_\nu\right)^2 - \sum_{i=x,y,z} \left(p_{\bar{\ell}_i} + p_{\nu_i}\right)^2 \tag{5.7}$$

$$m_{W^-}^2 = (E_\ell + E_{\bar{\nu}})^2 - \sum_{i=x,y,z} (p_{\ell_i} + p_{\bar{\nu}_i})^2 \tag{5.8}$$

Finally, the reconstructed top quark and anti-quark are assumed to have an invariant mass of $m_t = m_{\bar{t}} = 172.5\,\text{GeV}$:

$$m_t^2 = \left(E_{\bar{\ell}} + E_\nu + E_b\right)^2 - \sum_{i=x,y,z} \left(p_{\bar{\ell}_i} + p_{\nu_i} + p_{b_i}\right)^2 \tag{5.9}$$

$$m_{\bar{t}}^2 = \left(E_\ell + E_{\bar{\nu}} + E_{\bar{b}}\right)^2 - \sum_{i=x,y,z} \left(p_{\ell_i} + p_{\bar{\nu}_i} + p_{\bar{b}_i}\right)^2 \tag{5.10}$$

The constraint equations can be simplified to a single 4th order polynomial equation for one of the neutrino four-momenta components, which can be solved analytically with up to a four-fold ambiguity to obtain the neutrino and top quark four-momenta. If an event may contain more than two b-jet candidates, then each pair of jet–lepton assignments are tried. Jets b-tagged by the DeepJet algorithm are tried first, and only if no solution is found are the untagged jets considered. The solution which yields the minimum invariant mass of the $t\bar{t}$ system has been shown [13] to provide, in most cases, the solution with correct jet–lepton assignments and is the method used in this measurement to resolve the ambiguities arising from having multiple solutions.

Due to fluctuations in the measured jet momenta and missing transverse energy, the simplified constraint equation is not always solvable. To enhance reconstruction

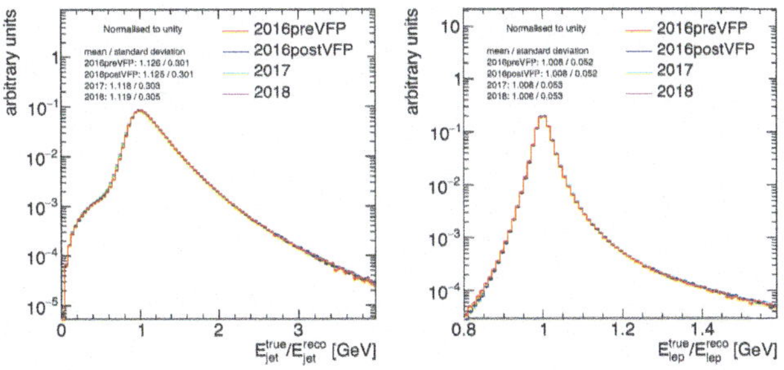

Fig. 5.8 Distributions of the true energy divided by the reconstructed energy used for the energy smearing in the $t\bar{t}$ full kinematic reconstruction. The factors are shown for the b quarks (left) and the leptons (right)

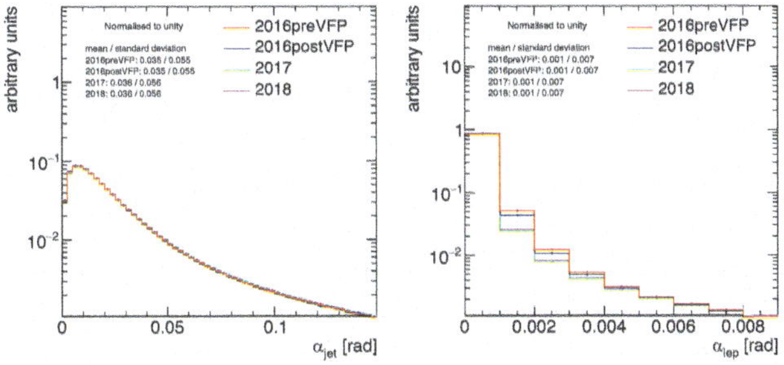

Fig. 5.9 Distributions of the angle between the true direction and the reconstructed direction for jets (left) and leptons (right)

efficiency, each event is reconstructed 100 times with different random smearings of jet and lepton energies and directions within their resolutions. The resolutions are determined using the signal MC simulation after event selection by comparing the reconstructed jet and lepton energies and directions to the true b-quark and lepton values. The jet and lepton energies are smeared by randomly sampling a distribution of the ratio of the true energy divided by the reconstructed energy (the ratio distributions are shown in Fig. 5.8). The directions are smeared randomly about the reconstructed direction with magnitude sampled from a Gaussian distribution with the resolution taken from the average of the angular difference between the reconstructed and true directions (the distributions from which the average resolutions were obtained are shown in Fig. 5.9). The W boson masses used in the constraints are also smeared randomly about their central values using the width of their relativistic Breit-Wigner mass distribution (shown in Fig. 5.10).

Likelihood weights for each smeared solution are calculated from the invariant mass of reconstructed leptons and b-jets relative to the true invariant mass distributions. The true lepton-b-jet invariant mass distribution used to calculate solution likelihood weights is shown in Fig. 5.10. For each smearing, the jet–lepton assignment that yields the maximum weight is chosen. A weighted average of all smeared solutions provides the final solution to the top quark three-momentum:

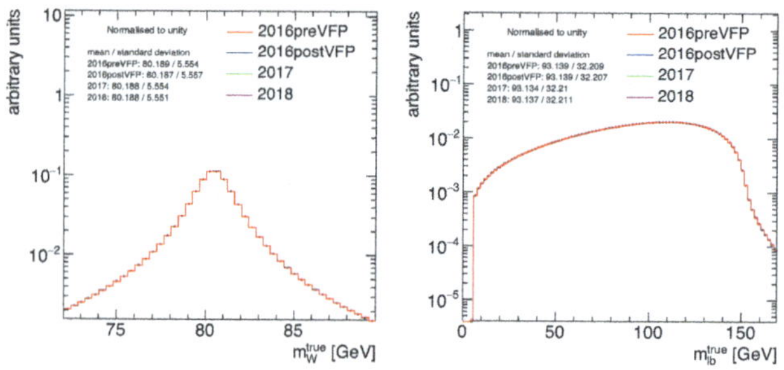

Fig. 5.10 Distributions of the true W boson mass (left) and invariant mass distribution of reconstructed leptons and b-jets (right)

$$\langle \vec{p}_t \rangle = \frac{1}{w(m_{\ell b})_\Sigma} \sum_i^{100} w(m_{\ell b})_i \cdot \vec{p}_{t_i} \tag{5.11}$$

where $w(m_{\ell b})_i$ and $\vec{p}_{t_i}$ denote the weight and reconstructed top quark three-momentum obtained for the i-th smearing of the event. Smearings without any solutions are given null weights. The top quark energy is calculated from its three-momentum and the top quark mass constraint $m_t = 172.5$ GeV. The final solution for the top anti-quark four-momentum is determined in an analogous procedure.

5.3.2 Kinematic Reconstruction Control Plots

Kinematic reconstruction control distributions for solutions to the kinematic reconstruction for combined 2016preVFP, 2016postVFP, 2017, 2018 eras and combined e^+e^-, $e^\pm\mu^\mp$, $\mu^+\mu^-$ channels after all object and event selection criteria have been applied are shown in Fig. 5.11 for b-jets, Fig. 5.12 for lepton and b-jet invariant mass, Fig. 5.13 for top quarks, and Fig. 5.14 for the $t\bar{t}$ system. The p_T spectra of top quarks and anti-quarks in $t\bar{t}$ data are known to be softer than modeled by MC simulations. Modeling systematics associated with this deficiency are discussed in Sect. 6.7.2 and it is also evident in the p_T distributions of downstream top quark decay products such as those for leptons (Fig. 5.7) and b-jets (Fig. 5.11).

5.3.3 Comparison of tt̄ Reconstruction for Prompt and Via τ Dilepton Events

The assumption that the dilepton decay is prompt and there are only two neutrinos in the final state is not true when the dilepton decay is via intermediate τ leptons;

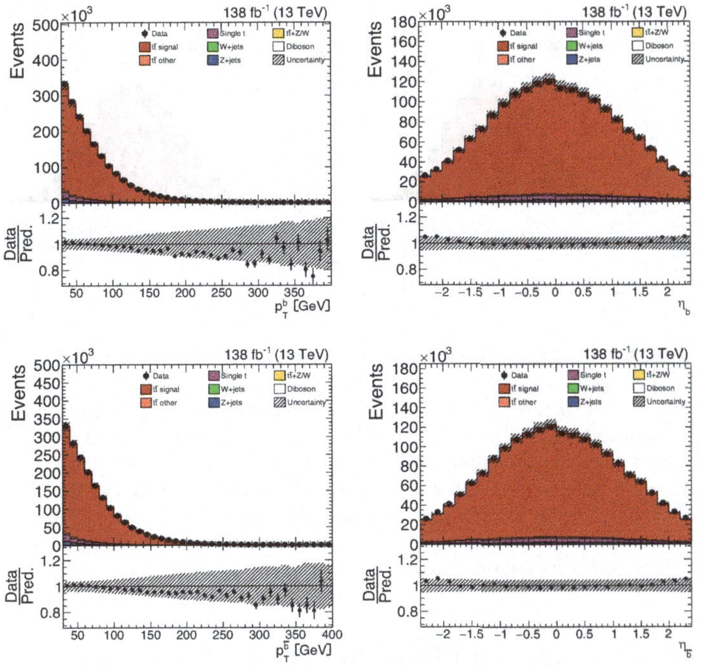

Fig. 5.11 p_{T} and η distributions for b-jet and anti-b-jet solutions to the kinematic reconstruction for combined 2016preVFP, 2016postVFP, 2017, 2018 eras and combined e^+e^-, $e^\pm\mu^\mp$, $\mu^+\mu^-$ channels after all object and event selection criteria have been applied. The simulated samples are re-weighted to an integrated luminosity of $137.7\mathrm{fb}^{-1}$. Trigger efficiency, lepton selection, b-tagging efficiency, and kinematic reconstruction efficiency scale factors discussed in Sect. 6.7.1 are applied. The Z+jets contributions in MC have been scaled to better describe the number of events observed in recorded data, as discussed in Sect. 5.4. Statistical uncertainties are shown as error bars on the data points, and total systematic uncertainties are indicated by the hatched uncertainty band

such decays can have four or six neutrinos in the final state. The top quark and $t\bar{t}$ reconstructed kinematic variables at detector-level (rec) are presented versus the true variables at generator level (gen), and additionally, the resolutions are presented as the difference between the true and reconstructed variables in bins of the true variables. Figures 5.15, 5.16, 5.17, 5.18, and 5.19 show the resolutions of $m_{t\bar{t}}$, $y_{t\bar{t}}$, ${p_{\mathrm{T}}}^{t\bar{t}}$, y_t, and ${p_{\mathrm{T}}}^t$ for both prompt and via tau reconstructed $t\bar{t}$ dilepton events. The means and standard deviation parameters (sigmas) from Gaussian fits to the kinematic variable resolutions are also included and reflect the bias and resolution of the algorithm, including degradation attributed to wrong jet–lepton assignments. From the sigmas of the fits, we conclude that the kinematic variable resolutions of $t\bar{t}$ dilepton events via τ are ~5–15% degraded relative to prompt events, but are sufficiently comparable to be included as signal in this measurement.

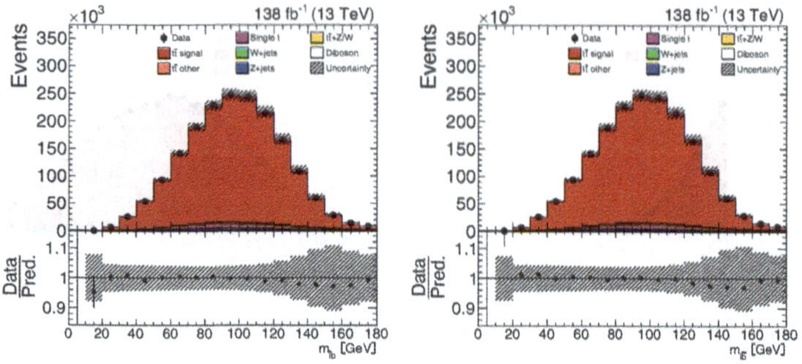

Fig. 5.12 Left (Right): Kinematic reconstruction solutions for anti-lepton and b-jet (lepton and anti-b-jet) invariant mass distributions for combined 2016preVFP, 2016postVFP, 2017, 2018 eras and combined e^+e^-, $e^\pm\mu^\mp$, $\mu^+\mu^-$ channels after all object and event selection criteria have been applied. The simulated samples are re-weighted to an integrated luminosity of 137.7fb^{-1}. Trigger efficiency, lepton selection, b-tagging efficiency, and kinematic reconstruction efficiency scale factors discussed in Sect. 6.7.1 are applied. The Z+jets contributions in MC have been scaled to better describe the number of events observed in recorded data, as discussed in Sect. 5.4. Statistical uncertainties are shown as error bars on the data points, and total systematic uncertainties are indicated by the hatched uncertainty band

5.4 Z+Jets Background Determination

The expected background contributions from $t\bar{t}$, tW, W+jets, and diboson processes are directly taken from the MC simulations. The Z+jets background contribution in the selected phase space is not modeled accurately in MC simulations, and the normalization of the Z+jets background contribution is corrected using global normalization scale factors, which are determined from a binned template fit to the data using TFractionFitter as described in [14, 15]. TFractionFitter is a class in the ROOT [16] data analysis framework that performs a simultaneous binned Poisson maximum likelihood fit of multiple MC templates in order to determine the linear combination of templates and fractions that best describe the recorded data. Within the vetoed Z boson mass peak control region $76\,\text{GeV} < m_{\ell\bar{\ell}} < 106\,\text{GeV}$, MC templates obtained from the MC simulated samples for Z+jets processes T_{Z+jets} and the sum of all other MC contributions T_{Other} are fitted to the $m_{\ell\bar{\ell}}$ distributions in the data template T_{Data} as:

$$\text{T}_{\text{Data}} = f_{Z+jets} \cdot \text{T}_{Z+jets} + f_{\text{Other}} \cdot \text{T}_{\text{Other}} \tag{5.12}$$

where f_{Z+jets} and f_{Other} are the fractions obtained from the fit. Separate global normalization scale factors (SF) are fitted in the e^+e^- and $\mu^+\mu^-$ channels:

$$SF_{ee} = \frac{f^{e^+e^-}_{Z+jets} \cdot N^{e^+e^-}_{\text{Data}}}{N^{e^+e^-}_{Z+jets}} \tag{5.13}$$

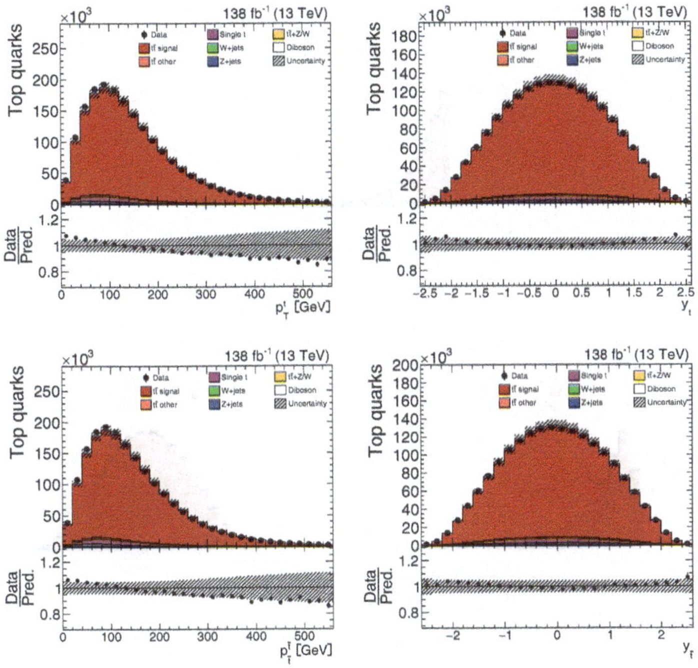

Fig. 5.13 p_T and y distributions for the top quark and anti-top quark solutions to the kinematic reconstruction for combined 2016preVFP, 2016postVFP, 2017, 2018 eras and combined e^+e^-, $e^\pm\mu^\mp$, $\mu^+\mu^-$ channels after all object and event selection criteria have been applied. The simulated samples are re-weighted to an integrated luminosity of 137.7fb^{-1}. Trigger efficiency, lepton selection, b-tagging efficiency, and kinematic reconstruction efficiency scale factors discussed in Sect. 6.7.1 are applied. The Z+jets contributions in MC have been scaled to better describe the number of events observed in recorded data, as discussed in Sect. 5.4. Statistical uncertainties are shown as error bars on the data points, and total systematic uncertainties are indicated by the hatched uncertainty band. The p_T spectra of top quark and anti-quarks in $t\bar{t}$ data are known to be softer than modeled by MC simulations, and modeling systematics associated with this deficiency are discussed in Sect. 6.7.2

$$SF_{\mu\mu} = \frac{f_{Z+jets}^{\mu^+\mu^-} \cdot N_{\text{Data}}^{\mu^+\mu^-}}{N_{Z+jets}^{\mu^+\mu^-}} \tag{5.14}$$

where N_{Z+jets} and N_{Data} are the number of events in the Z+jets template and data distribution. The $e^\pm\mu^\mp$ and combined channel scale factors are calculated as:

$$SF_{e^\pm\mu^\mp} = \sqrt{SF_{e^+e^-} \times SF_{\mu^+\mu^-}} \tag{5.15}$$

$$SF_{\ell\bar{\ell}} = \frac{SF_{e^+e^-} \cdot N_{\text{Data}}^{e^+e^-} + SF_{\mu^+\mu^-} \cdot N_{\text{Data}}^{\mu^+\mu^-} + SF_{e^\pm\mu^\mp} \cdot N_{\text{Data}}^{e^\pm\mu^\mp}}{N_{\text{Data}}^{e^+e^-} + N_{\text{Data}}^{\mu^+\mu^-} + N_{\text{Data}}^{e^\pm\mu^\mp}}. \tag{5.16}$$

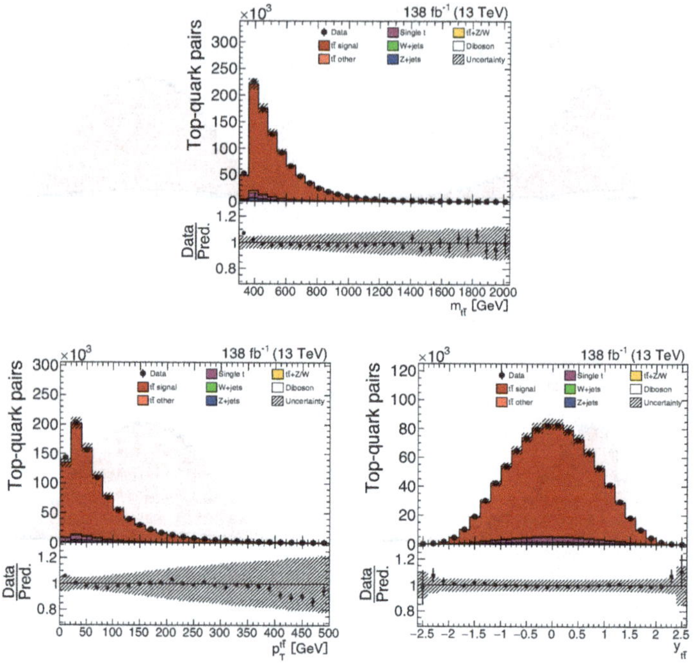

Fig. 5.14 p_T, y, and invariant mass distributions for $t\bar{t}$ solutions to the kinematic reconstruction for combined 2016preVFP, 2016postVFP, 2017, 2018 eras and combined e^+e^-, $e^\pm\mu^\mp$, $\mu^+\mu^-$ channels after all object and event selection criteria have been applied. The simulated samples are re-weighted to an integrated luminosity of 137.7fb^{-1}. Trigger efficiency, lepton selection, b-tagging efficiency, and kinematic reconstruction efficiency scale factors discussed in Sect. 6.7.1 are applied. The Z+jets contributions in MC have been scaled to better describe the number of events observed in recorded data, as discussed in Sect. 5.4. Statistical uncertainties are shown as error bars on the data points, and total systematic uncertainties are indicated by the hatched uncertainty band

To enhance Z+jets statistics in the orthogonal phase space of the Z boson mass peak control region, E_T^{miss} and b-tagging event selection requirements are omitted, as is the requirement of a kinematic reconstruction solution.

The fits performed by TFractionFitter of the two MC templates to the data template within the Z boson mass peak control region are shown for Full Run II in Fig. 5.20. The resulting Z+jets global normalization scale factors are shown in Table 5.2, and the dilepton invariant mass distribution after applying the scale factors is shown in Fig. 5.21.

5.5 Event Yields

The consecutive event selection steps are summarized as:

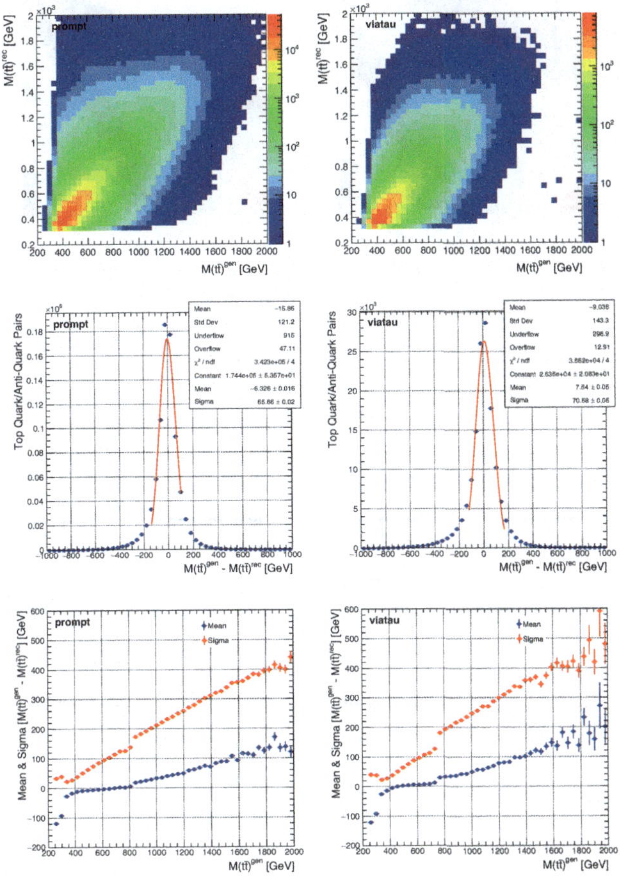

Fig. 5.15 Top: True and reconstructed $m_{t\bar{t}}$ obtained for prompt (left) and via tau (right) $t\bar{t}$ dilepton events. Middle: The difference between true and reconstructed $m_{t\bar{t}}$ in bins of true $m_{t\bar{t}}$ (fitted with a Gaussian function for illustration). Bottom: The differential mean and sigma from Gaussian fits, with respect to true $m_{t\bar{t}}$, of the difference between true and reconstructed $m_{t\bar{t}}$ obtained for prompt and via tau $t\bar{t}$ dilepton events. The simulated samples are normalized to an integrated luminosity of $137.7\,\text{fb}^{-1}$

- **MET Filters:** Events that are not excluded due to MET filters,
- **Triggers:** Events that pass trigger requirements,
- **Primary Vertex:** Events with primary vertex passing selection requirements,
- **Leptons:** Events with exactly two oppositely-charged isolated leptons and $m_{\ell\ell} > 20\,\text{GeV}$. In the $\mu^+\mu^-$ and e^+e^- channels, the Z+jets background is removed by rejecting events in the Z boson mass region $76\,\text{GeV} < m^{\ell\bar{\ell}} < 106\,\text{GeV}$,
- **Jets:** Events with at least two jets fulfilling the jet p_{T} and η requirements,
- **Missing Transverse Energy:** Events in the $\mu^+\mu^-$ and e^+e^- channels with $E_{\text{T}}^{\text{miss}} > 40\,\text{GeV}$,

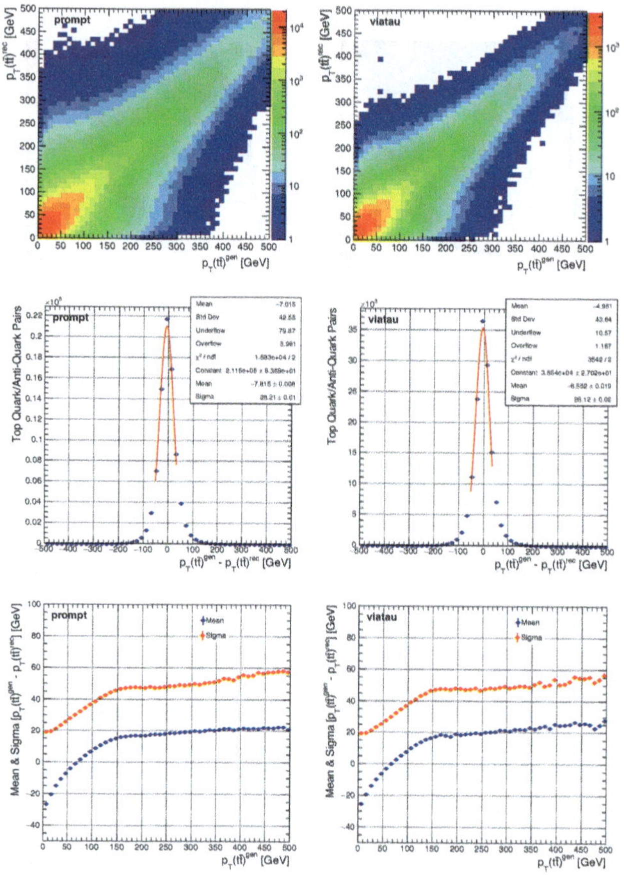

Fig. 5.16 Top: True and reconstructed $p_T{}^{t\bar{t}}$ obtained for prompt (left) and via tau (right) $t\bar{t}$ dilepton events. Middle: The difference between true and reconstructed $p_T{}^{t\bar{t}}$ in bins of true $p_T{}^{t\bar{t}}$ (fitted with a Gaussian function for illustration). Bottom: The differential mean and sigma from Gaussian fits, with respect to true $p_T{}^{t\bar{t}}$, of the difference between true and reconstructed $p_T{}^{t\bar{t}}$ obtained for prompt and via tau $t\bar{t}$ dilepton events. The simulated samples are normalized to an integrated luminosity of 137.7fb^{-1}

- ***b*-Tagging:** Events include at least one *b*-tagged jet,
- **Top Pair Reconstruction:** Events with at least one physically meaningful solution for the full kinematic reconstruction of a $t\bar{t}$ system.

The numbers of observed events in the data compared to the expected events from MC simulation for Full Run II are given in Table 5.3 after each of the last five consecutive selection steps.

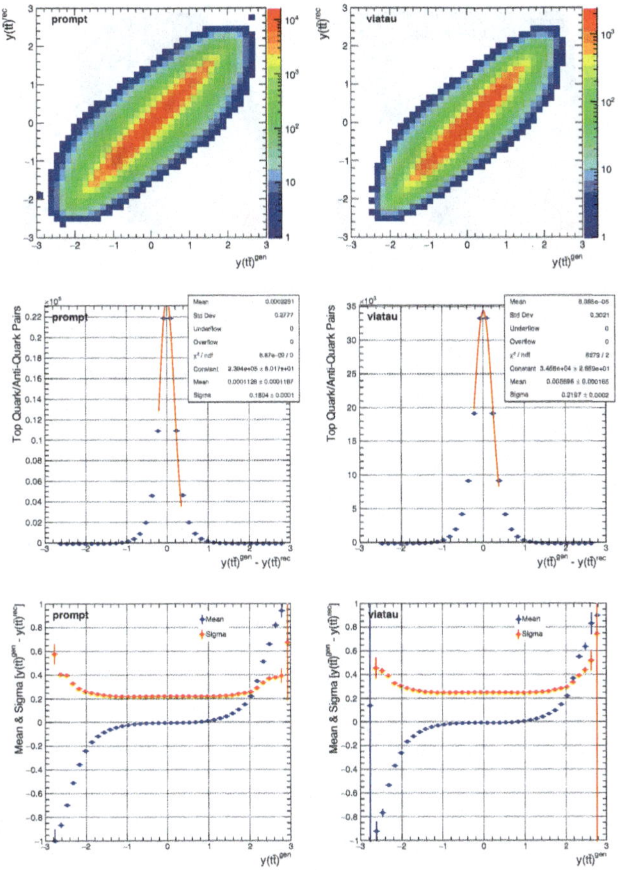

Fig. 5.17 Top: True and reconstructed $y_{t\bar{t}}$ obtained for prompt (left) and via tau (right) $t\bar{t}$ dilepton events. Middle: The difference between true and reconstructed $y_{t\bar{t}}$ in bins of true $y_{t\bar{t}}$ (fitted with a Gaussian function for illustration). Bottom: The differential mean and sigma from Gaussian fits, with respect to true $y_{t\bar{t}}$, of the difference between true and reconstructed $y_{t\bar{t}}$ obtained for prompt and via tau $t\bar{t}$ dilepton events. The simulated samples are normalized to an integrated luminosity of 137.7fb^{-1}

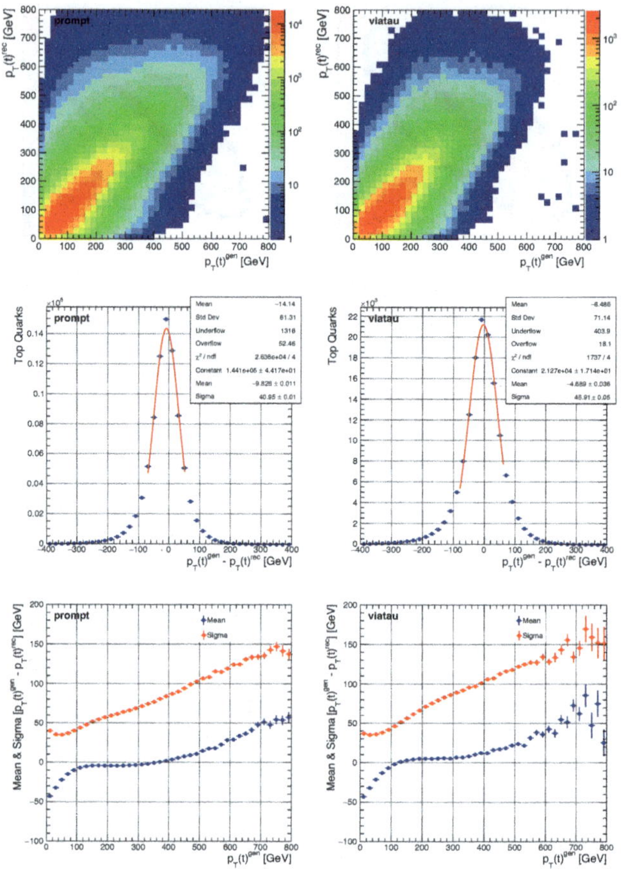

Fig. 5.18 Top: True and reconstructed $p_T{}^t$ obtained for prompt (left) and via tau (right) $t\bar{t}$ dilepton events. Middle: The difference between true and reconstructed $p_T{}^t$ in bins of true $p_T{}^t$ (fitted with a Gaussian function for illustration). Bottom: The differential mean and sigma from Gaussian fits, with respect to true $p_T{}^t$, of the difference between true and reconstructed $p_T{}^t$ obtained for prompt and via tau $t\bar{t}$ dilepton events. The simulated samples are normalized to an integrated luminosity of 137.7fb^{-1}

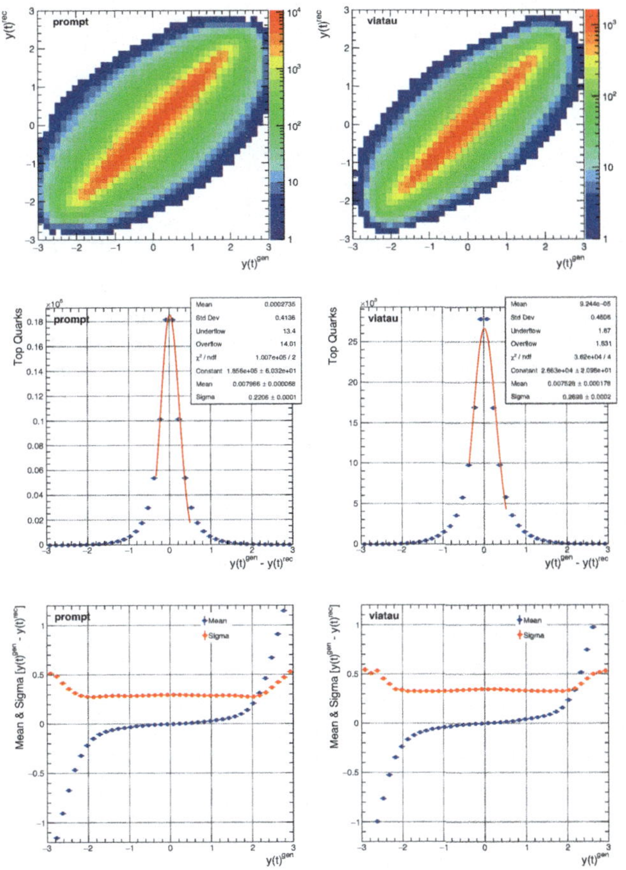

Fig. 5.19 Top: True and reconstructed y_t obtained for prompt (left) and via tau (right) $t\bar{t}$ dilepton events. Middle: The difference between true and reconstructed y_t in bins of true y_t (fitted with a Gaussian function for illustration). Bottom: The differential mean and sigma from Gaussian fits, with respect to true y_t, of the difference between true and reconstructed y_t obtained for prompt and via tau $t\bar{t}$ dilepton events. The simulated samples are normalized to an integrated luminosity of 137.7fb^{-1}

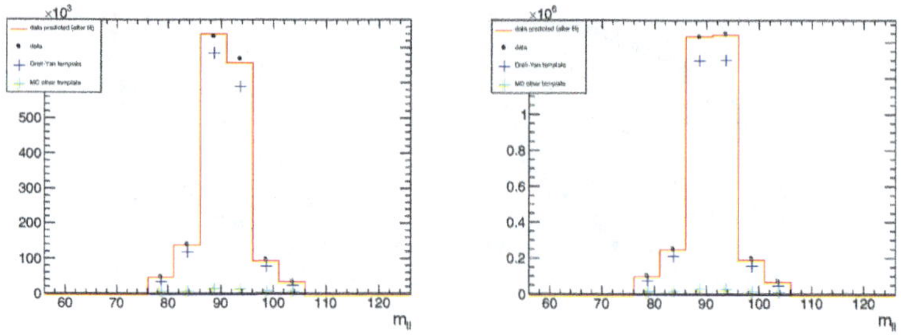

Fig. 5.20 Dilepton mass distributions in the Z boson mass peak control region between 76 GeV $< m_{\ell\bar{\ell}} <$ 106 GeV are shown for the e^+e^- (left) and $\mu^+\mu^-$ (right) channels for Full Run II. The data template $\mathrm{T_{Data}}$ is illustrated by black dots, the Z+jets template T_{Z+jets} is shown in blue, and the sum of all other MC contributions template $\mathrm{T_{Other}}$ is green. The red histogram shows the result of the fit performed by TFractionFitter

Table 5.2 Data-driven Z+jets global normalization scale factors ± the statistical uncertainty from the fit results, determined within the vetoed Z boson mass peak control region 76 GeV $< m_{\ell\bar{\ell}} <$ 106 GeV using TFractionFitter. The uncertainties correspond to statistical uncertainties for the fit results

Z+jets global normalization scale factors				
	$SF_{e^+e^-}$	$SF_{\mu^+\mu^-}$	$SF_{e^\pm\mu^\mp}$	$SF_{\ell\bar{\ell}}$
2016preVFP	1.081 ± 0.005	1.071 ± 0.004	1.076 ± 0.013	1.074 ± 0.019
2016postVFP	1.112 ± 0.006	1.144 ± 0.004	1.128 ± 0.015	1.133 ± 0.021
2017	1.078 ± 0.004	1.103 ± 0.002	1.091 ± 0.009	1.095 ± 0.012
2018	1.042 ± 0.003	1.047 ± 0.002	1.044 ± 0.007	1.045 ± 0.01
Full Run II	1.067 ± 0.002	1.079 ± 0.001	1.073 ± 0.005	1.075 ± 0.007

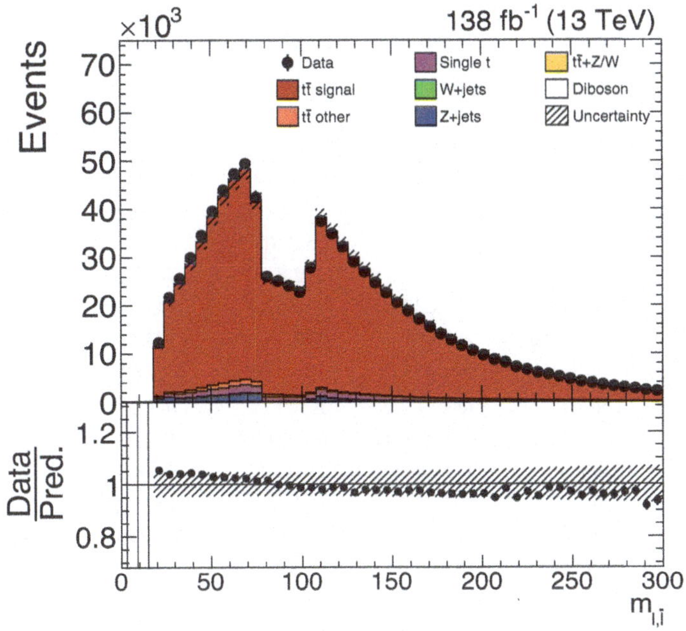

Fig. 5.21 Dilepton invariant mass distribution for combined 2016preVFP, 2016postVFP, 2017, 2018 eras and combined e^+e^-, $e^\pm\mu^\mp$, $\mu^+\mu^-$ channels after all object and event selection criteria have been applied, with Z+jets contributions in MC scaled to better describe the number of events observed in recorded data. The simulated samples are re-weighted to an integrated luminosity of 137.7fb^{-1}. Trigger efficiency, lepton selection, b-tagging efficiency, and kinematic reconstruction efficiency scale factors discussed in Sect. 6.7.1 are applied. Statistical uncertainties are shown as error bars on the data points, and total systematic uncertainties are indicated by the hatched uncertainty band

Table 5.3 Full Run II (137.7 fb^{-1}) expected event yields for signal and background processes, compared to the event yields in recorded data, after each step of event selection

$\mu^+\mu^-$	Leptons	Jets	E_T^{miss}	b-Tag	Kin. Reco.
$t\bar{t}$ Signal	486,260.0	342,459.0	272,219.0	241,382.0	220,154.0
$t\bar{t}$ Other	10,796.4	8293.4	6237.5	4266.4	3544.3
$t\bar{t}+Z/W$	1326.0	1201.9	988.9	850.3	664.0
Single	49,638.5	19,065.8	15,251.4	12,362.2	8413.4
Diboson	71,079.5	6822.1	3549.0	475.0	275.8
W+jets	5268.4	589.4	252.8	78.9	78.3
Z+jets	7,489,000.0	404,230.0	91,862.9	14,070.5	9951.6
MC Total	8,113,370.0	782,661.0	390,362.0	273,486.0	**243,081.0**
Data	8,060,660.0	797,354.0	405,879.0	278,054.0	**247,150.0**
$\mu^\pm e^\mp$	Leptons	Jets	E_T^{miss}	b-Tag	Kin. Reco.
$t\bar{t}$ Signal	896,402.0	633,218.0	633,218.0	562,837.0	520,593.0
$t\bar{t}$ Other	15,438.4	11,944.2	11,944.2	8448.5	7157.5
$t\bar{t}+Z/W$	2174.8	1954.3	1954.3	1676.3	1382.2
Single	90,803.8	34,783.6	34,783.6	28,290.6	20,105.7
Diboson	111,527.0	6568.6	6568.6	773.8	475.7
W+jets	17,722.3	1904.9	1904.9	354.6	248.2
Z+jets	269,816.0	17,614.8	17,614.8	2297.5	1803.2
MC Total	1403,880.0	707,988.0	707,988.0	604,679.0	**551,765.0**
Data	1,441,170.0	710,103.0	710,103.0	598,682.0	**546,299.0**
e^+e^-	Leptons	Jets	E_T^{miss}	b-Tag	Kin. Reco.
$t\bar{t}$ Signal	257,108.0	181,548.0	144,446.0	127,904.0	115,659.0
$t\bar{t}$ Other	2370.2	1855.0	1405.8	1083.1	943.4
$t\bar{t}+Z/W$	742.8	675.2	551.2	469.2	356.6
Single	26,190.4	10,211.3	8179.1	6673.7	4369.0
Diboson	35,619.4	3645.2	1920.4	264.2	150.9
W+jets	5859.0	454.1	321.0	13.3	0.0
Z+jets	3,282,680.0	191,128.0	42,368.6	6563.1	4360.6
MC Total	3,610,570.0	389,517.0	199,192.0	142,971.0	**125,840.0**
Data	3,535,030.0	388,972.0	201,114.0	140,293.0	**123,471.0**
$\ell\bar{\ell}$	Leptons	Jets	E_T^{miss}	b-Tag	Kin. Reco.
$t\bar{t}$ Signal	1,639,770.0	1,157,230.0	1,049,880.0	932,124.0	856,406.0
$t\bar{t}$ Other	28,605.1	22,092.6	19,587.5	13,797.9	11,645.2
$t\bar{t}+Z/W$	4243.7	3831.4	3494.4	2995.8	2402.8
Single	166,633.0	64,060.7	58,214.0	47,326.4	32,888.1
Diboson	218,226.0	17,035.9	12,038.0	1513.0	902.4
W+jets	28,849.8	2948.4	2478.7	446.8	326.5
Z+jets	11,041,500.0	612,934.0	151,855.0	22,932.3	16,114.4
MC Total	13,127,800.0	1,880,130.0	1,297,550.0	1,021,140.0	**920,685.0**
Data	13,036,800.0	1,896,430.0	1,317,100.0	1,017,030.0	**916,920.0**

References

1. CMS Collaboration, Precision luminosity measurement in proton-proton collisions at $\sqrt{s} = 13$ TeV in 2015 and 2016 at CMS. Eur. Phys. J. C **81**(9), 800 (2021)
2. CMS Collaboration, CMS luminosity measurement for the 2017 data-taking period at $\sqrt{s} =$ 13 TeV. CMS Physics Analysis Summary CMS-PAS-LUM-17-004 (2018). [Online]. Available: http://cds.cern.ch/record/2621960
3. CMS Collaboration, CMS luminosity measurement for the 2018 data-taking period at $\sqrt{s} =$ 13 TeV. CMS Physics Analysis Summary CMS-PAS-LUM-18-002 (2019). [Online]. Available: http://cds.cern.ch/record/2676164
4. CMS Collaboration, Particle-flow reconstruction and global event description with the CMS detector. Particle-flow reconstruction and global event description with the CMS detector. JINST **12**(10), P10003 (2017). Replaced with the published version. Added the journal reference and DOI. All the figures and tables can be found at http://cms-results.web.cern.ch/cms-results/public-results/publications/PRF-14-001 (CMS Public Pages). https://doi.org/10.1088/1748-0221/12/10/P10003. arXiv: 1706.04965. [Online]. Available: https://cds.cern.ch/record/2270046
5. CMS Collaboration, The CMS experiment at the CERN LHC. The Compact Muon Solenoid experiment. JINST **3**, S08004 (2008). Also published by CERN Geneva in 2010. https://doi.org/10.1088/1748-0221/3/08/S08004. [Online]. Available: http://cds.cern.ch/record/1129810
6. T. McCauley, Collisions recorded by the CMS detector on 14 Oct 2016 during the high pile-up fill. CMS Collection (2016). [Online]. Available: https://cds.cern.ch/record/2231915
7. CMS Collaboration, Pileup mitigation at CMS in 13 TeV data. JINST **15**(09), P09018 (2020)
8. M. Cacciari, G.P. Salam, G. Soyez, The anti-kt jet clustering algorithm. J. High Energy Phys. **2008**(04), 063 (2008). [Online]. Available: https://doi.org/10.1088/1126-6708/2008/04/063
9. T. Sakuma, Missing transverse energy diagram. [Online]. Available: https://cms-docdb.cern.ch/cgi-bin/PublicDocDB/ShowDocument?docid=12312
10. E. Bols, J. Kieseler, M. Verzetti, M. Stoye, A. Stakia, Jet flavour classification using DeepJet. J. Instrum. **15**(12), P12012 (2020). [Online]. Available: https://doi.org/10.1088/1748-0221/15/12/p12012.
11. A. Cagnotta, F. Carnevali, A. De Iorio, Machine learning applications for jet tagging in the cms experiment. Appl. Sci. **12**(20) (2022). ISSN: 2076–3417. https://doi.org/10.3390/app122010574. [Online]. Available: https://www.mdpi.com/2076-3417/12/20/10574
12. L. Sonnenschein, Analytical solution of ttbar dilepton equations. Phys. Rev. **D73**, 054015 (2006). https://doi.org/10.1103/PhysRevD.78.079902, https://doi.org/10.1103/PhysRevD.73.054015. arXiv: hep-ph/0603011 [hep-ph]
13. A. Abulencia et al., Measurement of the top quark mass using template methods on dilepton events in $p\bar{p}$ collisions at $\sqrt{s} = 1.96$ TeV. Phys. Rev. D **73**, 112006 (2006). [Online]. Available: https://doi.org/10.1103/PhysRevD.73.112006
14. CMS Collaboration, Measurement of differential cross sections for the production of top quark pairs and of additional jets in pp collisions at $\sqrt{s} = 13$ TeV. CERN, Geneva, Tech. Rep. (2022). [Online]. Available: http://cds.cern.ch/record/2803771
15. R. Barlow, C. Beeston, Fitting using finite monte carlo samples. Comput. Phys. Commun. **77**(2), 219–228 (1993). ISSN: 0010-4655. https://doi.org/10.1016/0010-4655(93)90005-W. [Online]. Available: https://www.sciencedirect.com/science/article/pii/001046559390005W
16. I. Antcheva et al., ROOT: a C++ framework for petabyte data storage, statistical analysis and visualization. Comput. Phys. Commun. **182**, 1384 (2011). https://doi.org/10.1016/j.cpc.2011.02.008

Chapter 6
Measurements of Differential Cross-Sections

Abstract This chapter discusses the procedures for measurement of differential cross-sections and the unfolding procedure used to correct for detector effects and obtain parton-level differential cross-sections extrapolated to the full phase space that are the focal points of the this dissertation. The sources of systematic uncertainties associated with these measurements are highlighted, as well as the methods employed to mitigate them.

The measurements of differential cross-sections presented in this dissertation follow the analysis strategy of CMS TOP-18-006 [1] but have been extended with additional observables, differential measurements as a function of $t\bar{t}$ invariant mass, several optimizations that reduce systematic uncertainties and significantly increased luminosity. All measured distributions are corrected for detector efficiencies and acceptances and extrapolated to parton-level using a regularized unfolding procedure with detector response obtained from MC simulated SM predictions with NLO matrix element accuracy interfaced with parton-shower algorithms. The spin density coefficients are extracted from the unfolded distributions, constituting a precision test of the SM.

6.1 Reconstructed Top Quark Polarization and $\mathbf{t\bar{t}}$ Spin Correlation Observables

Top quark polarization and $t\bar{t}$ spin correlation observables $\cos\theta^i_{1(2)}$, $\cos\theta^i_1\cos\theta^i_2$, and $\cos\theta^i_1\cos\theta^j_2 \pm \cos\theta^j_1\cos\theta^i_2$ are calculated using reconstructed top quarks and leptons as described in Sect. 3.4, where the quantity $\cos\theta^k_1 = \hat{\hat{\ell}}_1 \cdot \hat{k}$, as an example, can be calculated by taking the dot product of the charged anti-lepton direction in its grandfather top quark's ZMF with the $\hat{k}$ helicity basis spin quantization axis

J. R. Thieman, *Top Quark Spin Correlations with the CMS Detector*,
Springer Theses, https://doi.org/10.1007/978-3-031-76117-1_6

Table 6.1 A summary of all spin density observables, their respective coefficients, and which coefficient functions they probe

Observable	Measured coefficient	Coefficient function
$\cos\theta_1^k$	B_1^k	b_k^+
$\cos\theta_2^k$	B_2^k	b_k^-
$\cos\theta_1^r$	B_1^r	b_r^+
$\cos\theta_2^r$	B_2^r	b_r^-
$\cos\theta_1^n$	B_1^n	b_n^+
$\cos\theta_2^n$	B_2^n	b_n^-
$\cos\theta_1^k \cos\theta_2^k$	C_{kk}	c_{kk}
$\cos\theta_1^r \cos\theta_2^r$	C_{rr}	c_{rr}
$\cos\theta_1^n \cos\theta_2^n$	C_{nn}	c_{nn}
$\cos\theta_1^r \cos\theta_2^k + \cos\theta_1^k \cos\theta_2^r$	$C_{rk} + C_{kr}$	c_{rk}
$\cos\theta_1^r \cos\theta_2^k - \cos\theta_1^k \cos\theta_2^r$	$C_{rk} - C_{kr}$	c_n
$\cos\theta_1^n \cos\theta_2^r + \cos\theta_1^r \cos\theta_2^n$	$C_{nr} + C_{rn}$	c_{nr}
$\cos\theta_1^n \cos\theta_2^r - \cos\theta_1^r \cos\theta_2^n$	$C_{nr} - C_{rn}$	c_k
$\cos\theta_1^n \cos\theta_2^k + \cos\theta_1^k \cos\theta_2^n$	$C_{nk} + C_{kn}$	c_{kn}
$\cos\theta_1^n \cos\theta_2^k - \cos\theta_1^k \cos\theta_2^n$	$C_{nk} - C_{kn}$	$-c_r$
$\cos\varphi$	D	$-(c_{kk} + c_{rr} + c_{nn})/3$

in the $t\bar{t}$ ZMF. Additionally, $\cos\varphi = \hat{\hat{\ell}}_1 \cdot \hat{\ell}_2$, the dot product of the charged lepton directions in their grandfather top quark's ZMF, sensitive to coefficient $D = -\frac{1}{3}\mathrm{Tr}[\mathbf{C}] = -\frac{1}{3}(C_{kk} + C_{rr} + C_{nn})$ with single-differential cross-section $\frac{1}{\sigma}\frac{d\sigma}{d\cos\varphi} = \frac{1}{2}(1 - D\cos\varphi)$ is also measured. Table 6.1 provides a summary of all spin density observables, their respective coefficients, and which coefficient functions of the spin density matrix they probe. All measured distributions are measured both in the full selected phase space and differentially, with the phase space partitioned in bins of $t\bar{t}$ invariant mass $m_{t\bar{t}}$. The differential measurements are well-motivated based on the revelations discussed in Sect. 3.3, that the kinematic and spin properties of a $t\bar{t}$ pair can be completely described at LO in the $t\bar{t}$ ZMF by the invariant mass $m_{t\bar{t}}$ and the top scattering angle Θ.

6.2 Background Subtraction

Prior to unfolding, the predicted MC simulated signal and background rates are used to subtract the estimated background contributions in each bin of distributions measured with recorded data:

$$N^{Data}_{Signal} = N^{Data}_{Total} - N^{MC}_{Background} \tag{6.1}$$

Subtracting the estimated background from the measured data prior to unfolding reduces the impact of the unwanted background contributions and improves the accuracy of the unfolding procedure. Uncertainties on the subtracted background contributions are propagated to the remaining signal contributions.

6.3 Acceptance, Efficiency, Smearing, Purity, and Stability

The CMS detector has limited acceptance and reconstruction efficiency as well as finite resolution. Acceptance refers to the total number of events or particles that can be detected within the geometric coverage of the detector, and efficiency is the fraction of those events or particles that are reconstructed. The acceptance×efficiency is described by diagonal matrix **A**. The smearing of events between bins, due to finite detector resolution and imperfect reconstruction techniques, is described by matrix **S**. Matrices **A** and **S** can be quantified as:

$$\mathbf{A}_{\mathrm{ii}} = \frac{N_{\{x_i^{MC}\}\cap\{\vec{y}^{MC}\}}}{N_{\{x_i^{MC}\}}} \qquad \mathbf{S}_{\mathrm{ij}} = \frac{N_{\{y_j^{MC}\}}}{N_{\{x_i^{MC}\}\cap\{\vec{y}^{MC}\}}} \tag{6.2}$$

where $N_{\{x_i^{MC}\}\cap\{\vec{y}^{MC}\}}$ is the number of simulated events generated in the i-th bin that were reconstructed in any bin, $N_{\{x_i^{MC}\}}$ is the number of events generated in the i-th bin, and $N_{\{y_j^{MC}\}}$ is the number of simulated events reconstructed in the j-th bin.

The migration of events between bins due to smearing is also characterized by measures of purity of stability. Purity refers to the fraction of events in the reconstructed detector-level distribution that were actually generated in the same bin of the true generator-level distribution, with a high (low) purity for a bin indicating that a small (large) number of events in that bin of the reconstructed detector-level distribution migrated in from other bins in the true generator-level distribution. Stability refers to the fraction of events in a bin of the true generator-level distribution that were reconstructed in the same bin, with a high (low) stability indicating that a small (large) number of events in that bin of the true generator-level distribution migrated out to other bins of the reconstructed detector-level distribution. Purity and stability can be quantified as:

$$(\mathrm{Purity})_i = \frac{N_{\{x_i^{MC}\}\cap\{y_i^{MC}\}}}{N_{\{y_i^{MC}\}}} \qquad (\mathrm{Stability})_i = \frac{N_{\{x_i^{MC}\}\cap\{y_i^{MC}\}}}{N_{\{x_i^{MC}\}\cap\{\vec{y}^{MC}\}}} \tag{6.3}$$

where $N_{\{y_i^{MC}\}}$ is the number of simulated events reconstructed in the i-th bin, $N_{\{x_i^{MC}\}\cap\{\vec{y}^{MC}\}}$ is the number of events generated in the i-th bin that were reconstructed in any bin, and $N_{\{x_i^{MC}\}\cap\{y_i^{MC}\}}$ is the number of simulated events generated in the i-th bin that were also reconstructed in the i-th bin.

6.4 Unfolding Procedure

The event selection criteria and analysis techniques such as the $t\bar{t}$ kinematic reconstruction trade acceptance and efficiency in exchange for high purity of signal events. As a result, measured bins of reconstructed detector-level distributions $\vec{y}$ are distorted from the true generator-level distributions $\vec{x}$ by acceptance, efficiency, and bin-to-bin migrations:

$$\vec{y}^{\text{MC}} = \mathbf{M} \cdot \vec{x}^{\text{MC}} \qquad \xrightarrow[\text{Unfolding}]{\text{Naive}} \qquad \vec{x} = \mathbf{M}^{-1}\vec{y} \tag{6.4}$$

where $\mathbf{M}$ is the response-matrix constructed from $\mathbf{A}$ and $\mathbf{S}$ using signal process events from the POWHEGV2 + PYTHIA8 MC simulation including smearing from the GEANT4-simulated detector response. For each 1D (2D) measured observable, the response-matrix is a 2(4)-dimensional, reconstructed detector-level versus true generator-level distribution containing events that pass the event selection. In each generator-level bin, the true generated events in that bin that do not pass the event selection are filled in the underflow. To correct the bin of background-subtracted recorded data distributions for smearing and acceptance effects, the TUnfold [2] package of the ROOT data analysis framework is used to simultaneously invert $\mathbf{M}$ and unfold the detector-level distributions $\vec{y}$ of recorded data into distributions $\vec{x}$ that have been corrected for bin migrations and extrapolated to parton-level. For each measured distribution, the absolute differential cross-section, the differential cross-section normalized to unit area, and the extracted spin-density coefficients constitute the final results of the unfolding procedure.

6.4.1 Regularization

Regularized unfolding is used to suppress statistical fluctuations and reduce the impact of noise by introducing a term into the unfolding algorithm that incorporates assumptions that constrain the unfolded result to be smoother. This technique helps prevent over-fitting data and improves the stability of the unfolded result. The unfolding for these measurements is performed using the TUnfoldDensity class [2], which offers greater control over the regularization, using a χ^2 fit to simultaneously unfold $\vec{y}$ and regularize the unfolded result:

$$\chi^2_{\text{unf}} = \chi^2_{\mathbf{M}} + \tau^2\chi^2_{\mathbf{L}} + \lambda\sum_i(\mathbf{M}\vec{x} - \vec{y})_i \tag{6.5}$$

$$\text{where} \quad \chi^2_{\mathbf{M}} = (\mathbf{M}\vec{x} - \vec{y})^T\mathbf{V}_y^{-1}(\mathbf{M}\vec{x} - \vec{y}) \tag{6.6}$$

$$\chi^2_{\mathbf{L}} = (\vec{z})^T\mathbf{L}^T\mathbf{L}(\vec{z}) \quad \text{where} \quad z_i = f_i \cdot (x_i - \frac{\sum_i x_i}{\sum_i x_i^{\text{MC}}} \cdot x_i^{\text{MC}}) \tag{6.7}$$

where $\mathbf{V}_y$ is the diagonal covariance matrix of the reconstructed detector-level distribution $\vec{y}$ and $\mathbf{L}$ is a matrix specifying the regularization conditions. The $\chi^2_{\mathbf{M}}$ term is a least squares fit of the re-folded output $\mathbf{M}\vec{x}$ to the original data $\vec{y}$. In the absence of the other terms, the unfolded result will often have unacceptably large statistical fluctuations arising from strong anti-correlations between adjacent bins, and thus the expected shape $\vec{x}_{\text{MC}}$ of the distribution is used to regularize the output via the $\chi^2_{\mathbf{L}}$ term, with the strength of the regularization controlled by the τ parameter. Larger (smaller) values of τ mean the unfolded output will be over-smoothed (noisy) and more (less) strongly constrained by the generator-level MC vector $\vec{x}_{\text{MC}}$). For each unfolded result, the value of τ is chosen by scanning for the τ that yields the minimum of ρ_{avg}, the mean of the global correlation coefficient ρ_i in the bins of the unfolded distribution:

$$\rho_i = \sqrt{1 - \frac{1}{[(\mathbf{V}_x))_{ii} \cdot (\mathbf{V}_x)^{-1})_{ii}]}} \quad \text{with} \quad 0 \le \rho_i \le 1 \tag{6.8}$$

where $\mathbf{V}_x$ is the covariance matrix of the unfolded result $\vec{x}$ obtained through standard propagation of errors in the linear transformation of the reconstructed distribution $\vec{y}$ into $\vec{x}$, specified in Eq. 6.7. The $\lambda\Sigma_i(\mathbf{M}\vec{x} - \vec{y})_i$ term is a Lagrange multiplier that ensures the integral of the re-folded output matches that of the input.

For the unfolded measurements presented in this dissertation, the $\mathbf{L}$ matrix is chosen such that the regularization minimizes the curvature of the difference between the unfolded distribution $\vec{x}$ and the generator-level MC distribution $\frac{\sum_i x_i}{\sum_i x_i^{\text{MC}}} \cdot \vec{x}_{\text{MC}}$, where $\frac{\sum_i x_i}{\sum_i x_i^{\text{MC}}}$ is a scale factor that normalizes the total number of reconstructed-level MC signal events to the number observed in recorded data. Curvature regularization is unbiased when the shape of $\vec{x}$ can only differ from the shape of the MC simulation $\frac{\sum_i x_i}{\sum_i x_i^{\text{MC}}} \cdot \vec{x}_{\text{MC}}$ by a linear function of the measured variable. This is true for observables such as $\cos\theta$, where the SM analytical form of the differential cross-section is known to depend linearly on the measured coefficient B: $\frac{1}{\sigma}\frac{d\sigma}{d\cos\theta} = \frac{1}{2}(1 + B\cos\theta)$.

In order for the curvature matrix $\mathbf{L}$ to provide unbiased regularization for observables with differential cross-sections that have non-linear dependence, TUnfoldDensity allows for the inclusion of bin factor functions in the regularization term, with an appropriate f_i calculated for each bin such that $f_i \cdot (x_i - \frac{\sum_i x_i}{\sum_i x_i^{\text{MC}}} \cdot x_i^{\text{MC}})$ depend linearly on the measured coefficient. The function $f(x)$ is calculated with $g(x) - g_{\text{MC}}(x)$, the continuous analog of $(x_i - \frac{\sum_i x_i}{\sum_i x_i^{\text{MC}}} \cdot x_i^{\text{MC}})$, such that $z(x) = f(x) \cdot (g(x) - g_{\text{MC}}(x)) = \text{constant} \times x$, where $g(x)$ is the functional form of the unfolded distribution and $g_{\text{MC}}(x)$ is the functional form of the generated-level distribution in MC simulation (which assumes the SM value of the specific spin-density coefficient). The functional form of the distribution $g(x) = \frac{1}{\sigma}\frac{d\sigma}{dx}$ is known (see Sect. 3.4) for each of the observables measured in the top rest frame:

$$x = \cos\theta_1^i,\quad g(x) = \frac{1}{\sigma}\frac{d\sigma}{dx} = \frac{1}{2}(1 + B_1^i x) \tag{6.9}$$

$$x = \cos\theta_1^i \cos\theta_2^i,\quad g(x) = \frac{1}{\sigma}\frac{d\sigma}{dx} = \frac{1}{2}(1 - C_{ii}x)\log\left(\frac{1}{|x|}\right) \tag{6.10}$$

$$\begin{aligned} x &= \cos\theta_1^i \cos\theta_2^j \pm \cos\theta_1^j \cos\theta_2^i,\quad g(x) \\ &= \frac{1}{\sigma}\frac{d\sigma}{dx} = \frac{1}{2}\left(1 - \frac{C_{ij} \pm C_{ji}}{2}x\right)\cos^{-1}|x| \end{aligned} \tag{6.11}$$

Then, $f(x) \propto \frac{x}{g(x) - g_{\mathrm{MC}}(x)}$ follows:

$$x = \cos\theta_1^i,\quad f(x) = 1 \tag{6.12}$$

$$x = \cos\theta_1^i \cos\theta_2^i,\quad f(x) = \frac{1}{\log\left(\frac{1}{|x|}\right)} \tag{6.13}$$

$$x = \cos\theta_1^i \cos\theta_2^j \pm \cos\theta_1^j \cos\theta_2^i,\quad f(x) = \frac{1}{\cos^{-1}|x|} \tag{6.14}$$

All the distributions $g(x)$ considered are parameterized by only a single coefficient, so $f(x)$ never has any dependence on any model parameters and the regularization is always unbiased, even if NP contributions are present in the recorded data. The choice of the constant factor for $f(x)$ is inconsequential, as it is reabsorbed into the value of τ, and is arbitrarily chosen such that the average value of the distribution is unity. To apply the derived continuous bin factor functions to the binned distributions, $f(x)$ must be integrated over the boundaries of each bin to calculate the appropriate f_i for that bin.

For the differential unfolding of the observables as a function of $m_{t\bar{t}}$, TUnfold "unwraps" the two-dimensional distributions into one-dimensional distributions, and the unfolding procedure is performed similarly to as previously described. TUnfold allows regularization along the axis of the spin-density observable to prevent spurious statistical fluctuations as previously discussed, while not regularizing along the $m_{t\bar{t}}$ axis, as its functional form is not known.

6.4.2 *Binning*

Binning schemes for all distributions and matrices need to be chosen such that the resolution of the measured quantities are comparable to the bin widths. A binning scheme of six uniform bins $[-1, -\frac{2}{3}, -\frac{1}{3}, 0, +\frac{1}{3}, +\frac{2}{3}, +1]$ is shown to match the measurement resolution for all spin density coefficient distributions, and for differential measurements in $m_{t\bar{t}}$, four bins are used to partition the phase space [250, 450, 600, 800, 2000] GeV. To avoid having zero degrees of freedom in the least-squares term of χ^2_{unf}, reconstructed detector-level distributions are required to have twice as many bins as the true generator-level distributions.

With any binning scheme, information about the dependence of acceptance, efficiency, and migration inside each bin is lost, so the unfolding will be inherently biased. By unfolding with a finer binning scheme, more information is retained, the bias can be mitigated, and the precision improved. The unbiased regularization method means finer bins can be used without concern that the explicit regularization of the curvature might cause more bias than the implicit regularization from binning. To assess this, linearity tests are performed by injecting the distributions with changes in the spin-density coefficients by reweighting the simulated MC events and measuring the response of the unfolded results. Finer binning by a factor of four ($24 = 6 \cdot 4$ bins) is found to have near-perfect linearity and remove most of the implicit bias for one-dimensional unfolding, while for two-dimensional unfolding finer binning by a factor of two ($12 \times 8 = 6 \cdot 2 \times 4 \cdot 2$ bins) along each axis is found to be a good trade-off between removing bias and over-binning. To prevent the measurement sensitivity to features in the data smaller than the experimental resolution, the output distribution is re-binned immediately after the χ^2 fit and τ scan is performed with the original binning.

6.5 Conversion to Differential Cross-Sections

The i-th bin of the absolute differential cross-sections is obtained from the unfolded distributions $\vec{x}$ as:

$$\left(\frac{d\sigma}{dx}\right)_i = \frac{1}{\Delta x_i} \frac{x_i}{\mathcal{BF} \cdot \mathcal{L}} \tag{6.15}$$

where Δx_i is the bin width, $\mathcal{BF}$ is the branching fraction of the $t\bar{t}$ dilepton decay mode, and $\mathcal{L}$ is the integrated luminosity. Dividing by the total dileptonic $t\bar{t}$ production cross-section $\sigma = (\sum_i x_i)/(\mathcal{BF} \cdot \mathcal{L})$ normalizes the absolute differential cross-sections to unity:

$$\left(\frac{1}{\sigma}\frac{d\sigma}{dx}\right)_i = \frac{1}{\Delta x_i} \frac{x_i}{\sum_i x_i} \tag{6.16}$$

which is independent of $\mathcal{BF}$ and $\mathcal{L}$ (as these factors only affect the total rate of the process and not the shape of the distribution) thus eliminating systematic uncertainties due to these sources.

6.6 Extraction of Coefficients from Differential Cross-Sections

The coefficients C that parameterize the single-differential cross-sections:

$$g(x; C) = (1 \pm Cx)g(x; 0) \tag{6.17}$$

are extracted from asymmetries statistics A_i between bins i and $b+1-i$, defined as:

$$A_i = \frac{N_{b+1-i} - N_i}{N_{b+1-i} + N_i} \quad \text{for} \quad i = 1, 2, .., b/2 \tag{6.18}$$

where for these measurements $b = 6$ is the number of bins for all single-differential cross-sections. Using the known analytic form of the distribution, as well as the property that the coefficient induces an asymmetry between bins i and $b+1-i$, A_i can be written as:

$$A_i = \frac{\int_{bin}(1 \pm C_i x)g(x; 0)\, dx - \int_{bin}(1 \mp C_i x)g(x; 0)\, dx}{\int_{bin}(1 \pm C_i x)g(x; 0)\, dx + \int_{bin}(1 \mp C_i x)g(x; 0)\, dx} \tag{6.19}$$

$$\text{which implies} \quad C_i = \pm A_i \frac{\int_{bin} g(x; 0)\, dx}{\int_{bin} xg(x; 0)\, dx}. \tag{6.20}$$

where C_i is the coefficient extracted from the asymmetry between bins i and $b+1-i$, and the integrals over the bin region $\int_{bin} g(x; 0)\, dx$ and $\int_{bin} xg(x; 0)\, dx$ are evaluated using Mathematica. The C_i measurements, with covariance matrix V^C, can be combined for an optimal weighted measurement of C:

$$C = \frac{\sum_i C_i a_i}{\sum_i a_i} \qquad \sigma_C^2 = \frac{\vec{a}^T V^C \vec{a}}{(\sum_i a_i)^2}$$

with weights a_i determined by minimizing the variance of the combined measurement σ_C^2:

$$a_1 = \frac{b_1}{b_1 + b_2 + b_3}$$

$$b_1 = V^C_{2,2}V^C_{3,3} - V^C_{2,3}V^C_{2,3} - V^C_{1,2}(V^C_{3,3} - V^C_{2,3}) - V^C_{1,3}(V^C_{2,2} - V^C_{2,3})$$

$$a_2 = \frac{b_2}{b_1 + b_2 + b_3} \text{where}$$

$$b_2 = V^C_{1,1}V^C_{3,3} - V^C_{1,3}V^C_{1,3} - V^C_{1,2}(V^C_{3,3} - V^C_{1,3}) - V^C_{2,3}(V^C_{1,1} - V^C_{1,3})$$

$$a_3 = \frac{b_3}{b_1 + b_2 + b_3}$$

$$b_3 = V^C_{1,1}V^C_{2,2} - V^C_{1,2}V^C_{1,2} - V^C_{1,3}(V^C_{2,2} - V^C_{1,2}) - V^C_{2,3}(V^C_{1,1} - V^C_{1,2})$$

For small coefficients, these statistics can be shown to satisfy the Cramér-Rao bound and provide the minimum variances for unbiased estimators of the coefficients. For double differential measurements as a function of $m_{t\bar{t}}$, the coefficient extraction procedure is performed in an identical way in each of the sub-phase space regions partitioned by the $m_{t\bar{t}}$ bin boundaries.

6.7 Systematic Uncertainties

Each systematic uncertainty is evaluated in every measurement bin by repeating the full analysis with up and down $\pm\,1\sigma$ variations of a corresponding input parameter or event weight correction factor. For each systematic source, the difference between the varied result and the nominal result provides the respective uncertainty. For systematics that have many highly-correlated contributions, the combined impact of all contributions is obtained by enveloping the largest difference between the nominal value and varied results in each bin. The total up and down systematic uncertainties in each bin is then obtained by adding the uncorrelated systematic uncertainties in quadrature. The average of the total up and down variations provides the final estimate of the total systematic uncertainty for that bin.

6.7.1 Experimental Systematics

Experimental systematic uncertainties arise from detector imperfections and limitations of data analysis techniques. Sources of experimental systematic uncertainties include detector response, object reconstruction, efficiencies, background modeling, calibrations, and alignment.

- **Luminosity**
  The measurement uncertainties of the integrated luminosity delivered to the CMS detector during the 2016preVFP, 2016postVFP, 2017, and 2018 data-taking periods are 1.2% [3], 1.2% [3], 2.3% [4] and 2.5% [5] respectively. For the Full Run II combined measurement the uncertainty on the integrated luminosity is 1.6%. This uncertainty is a flat percentage in all measured bins, only impacts normalization and not shape, and cancels out in the measurements of normalized differential cross sections.

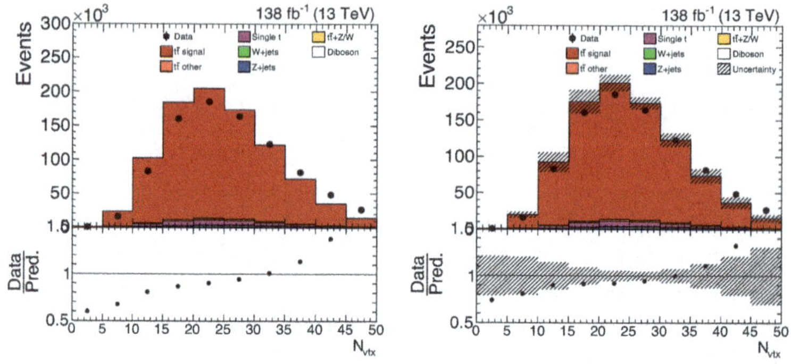

Fig. 6.1 Number of primary vertices distributions before (Left) and after (Right) PU re-weighting, for combined 2016preVFP, 2016postVFP, 2017, 2018 eras and combined e^+e^-, $e^\pm\mu^\mp$, $\mu^+\mu^-$ channels after all object and event selection criteria have been applied. The simulated samples are re-weighted to an integrated luminosity of 137.7fb^{-1}

- **Pileup**
  PU reweighting is performed using the instantaneous luminosity per bunch crossing in data, and the total pp inelastic cross section of 69.2 mb, to correct the PU distribution of the MC simulation. The effect of PU re-weighting is shown in Fig. 6.1. To estimate the systematic uncertainty due to PU modeling, the measurement is repeated with variations of $\pm$4.6% on the total pp inelastic cross-section input for PU reweighting.
- **Trigger Efficiency**
  The trigger efficiencies are measured two-dimensionally, as a function of p_T for both leptons, using E_T^{miss}-based triggers that are effectively uncorrelated with the dilepton triggers used in the measurement. The trigger efficiencies measured in data are used to correct the trigger efficiencies in MC simulations via corrective SFs that are typically close to unity. Event topology uncertainties for corrective SFs are estimated by recalculating the SFs in different kinematic phase space partitions: high (low) number of reconstructed jets and high (low) number of reconstructed vertices. Time-dependent uncertainties are estimated by recalculating the SFs in each of the run periods. The total trigger SF uncertainty is the quadrature sum of the event topology, run-dependency, and statistical sources. The measurement is repeated with the trigger SFs varied up and down $\pm$ 1σ. More information pertaining to trigger SFs can be found in Appendix A.
- **L1 ECAL and Muon "Prefiring"**
  In 2016 and 2017, a gradual timing shift of the ECAL was not properly propagated to the L1 trigger primitives, leading to an incorrect association of a significant fraction of high-$|\eta|$ trigger primitives to the previous bunch crossing. As L1 forbids firing of two consecutive bunch crossings, events can self-veto due to this effect if a significant amount of ECAL energy is found in the region of $2 < |\eta| < 3$ (known as L1 ECAL prefiring). The effect is strongly p_T and η-dependent. A similar effect is present in the muon system, where the

BX assignment of the muon candidates can be wrong due to the limited time resolution of the muon detectors. The L1 prefiring effect is not modeled in MC simulations must be accounted for by reweighting events. L1 prefiring weights are computed as the product of the non-prefiring probability of all objects present in the event:

$$w_{\text{L1 Prefiring}} = 1 - P(\text{Prefiring}) = \prod_{i=\text{photons, jets, muons}} (1 - \mathcal{E}_i^{\text{Prefiring}}(p_\text{T}, \eta)) \tag{6.21}$$

For 2016 and 2017 MC samples, the probability of both the ECAL and muon prefiring is used to calculate the event weights. However, for the 2018 MC samples, only the probability of Muon prefiring is used.

- **Lepton Selection Efficiencies and Energy Corrections**
  The identification and isolation efficiencies for muons or identification and reconstruction efficiencies of electrons are estimated using a tag-and-probe method with Z boson event samples as a function of p_T and η. Muon efficiencies are well-described in the simulation, with SFs in the range [0.95, 1.05], while for the electrons the SFs are typically in the range [0.9, 1.1]. SFs are varied individually for each lepton type and each SF component (i.e. identification/isolation for muons and identification+isolation/reconstruction for electrons). To account for the potential differences in isolation properties of leptons in the $t\bar{t}$ and Z+jets topologies, additional uncertainties of 0.5% for muons and 1% for electrons are assigned to the isolation component of scale factors. Furthermore, momentum scale and resolution corrections for muons and electrons are varied within their individual systematic uncertainties. The measurement is repeated with the varied four-vectors to determine the systematic uncertainty.
- **Jet Energy Scale**
  To determine the uncertainty due to JES corrections, the p_T and η-dependent JES corrections are varied within their uncertainties by $\pm\ 1\sigma$. For each source of JES uncertainty, the difference between the results obtained using the rescaled energies and the nominal results is taken as the systematic uncertainty. The total JES uncertainty is obtained via a quadratic sum of all sources of JES uncertainty.
- **Jet Energy Resolution**
  Corrections for JER are applied by smearing the jets in MC simulation to match the precision of jets in recorded data. When a matching true jet is found, reconstructed jet four-momenta are corrected with a p_T dependent SF. Otherwise, stochastic smearing is applied that only allows degradation of the resolution. The uncertainty due to JER corrections is determined by a variation of the η-dependent JER corrections to jets in the simulated samples by $\pm 1\sigma$.
- **Unclustered Missing Transverse Energy**
  The uncertainties on all the individual components contributing to the E_T^miss (like jets, muons, electrons, unclustered energy, etc.) are propagated to the MET, and changes in the E_T^miss four-vector caused by variations in the jet and lepton energy scales are accounted for in the corresponding uncertainties for jets and leptons.

This uncertainty includes only the variation due to the unclustered energy, which is accounted for by varying the deposited energy from the charged hadrons, neutral hadrons, and photons according to the corresponding energy resolutions and recalculating the $E_{\mathrm{T}}^{\mathrm{miss}}$ four-vector. The recalculated $E_{\mathrm{T}}^{\mathrm{miss}}$ is propagated through the measurement, and the difference of the variation from the nominal is taken as the systematic uncertainty.

- ***b*-Tagging**
  Due to the differences between b-tagging algorithm efficiencies in recorded data and MC simulation, MC simulated events are re-weighted with corrective scale factors after selection. Data-to-simulation corrective scale factors ($\mathrm{SF_{BTV}}$) for the b-tagging efficiency of individual b-jets, and mistag rate c-jets and light-(l) jets, are measured with an orthogonal QCD multi-jet dataset and parameterized as a function of the jet p_{T}. To correct the possible differences of the b-tagging efficiency due to the different kinematics of the $t\bar{t}$ events and the QCD multijet events used to measure the data-to-simulation corrective scale factors, the b-tagging efficiency ($\mathcal{E}_{\mathrm{MC}}$) of b-jets and mistag rate for c- and l-jets, which depend on both the jet kinematic properties and the event selection, are estimated in this analysis using the MC simulated $t\bar{t}$ signal events. The probability P of a given configuration of jets in MC simulation and data is defined as:

$$P(\mathrm{MC}) = \prod_{i=\mathrm{tagged}} \mathcal{E}_i \prod_{j=\mathrm{not\ tagged}} (1 - \mathcal{E}_j) \tag{6.22}$$

$$P(\mathrm{Data}) = \prod_{i=\mathrm{tagged}} SF_i\mathcal{E}_i \prod_{j=\mathrm{not\ tagged}} (1 - SF_j\mathcal{E}_j), \tag{6.23}$$

  where $\mathcal{E}_i$ and SF_i refer respectively to $\mathcal{E}_{\mathrm{MC}}$ and $\mathrm{SF_{BTV}}$, which are functions of the jet flavor, jet p_{T}, and jet η. Afterwards, the event weight is computed accordingly to $w_{\text{b-tag}} = \frac{P(\mathrm{Data})}{P(\mathrm{MC})}$.
  Depending on the flavor of the original parton that originated the jet shower, the measurement is repeated with the corrective SFs varied by $\pm\ 1\sigma$. The heavy flavor (b and c) jet SFs are varied separately from the light flavor jets.
- **Kinematic Reconstruction Efficiency**
  For a fraction of events ($\sim$10%), the kinematic reconstruction algorithm does not find a solution, and the events fail event selection. The kinematic reconstruction efficiency is measured in both MC simulated and recorded data from the ratio of events before and after the kinematic reconstruction and is used to derive global corrective scale factors for MC simulated data. The uncertainty of these scale factors is negligible and is not propagated to the total systematic uncertainty.
- **Backgrounds**
  While all sources of systematics are considered for the signal, the uncertainty due to background normalization is determined by a flat percentage variation of the background event counts. The background from Z+jets processes is varied by $\pm$30% in the e^+e^-, $e^\pm\mu^\mp$ and $\mu^+\mu^-$ channels, which is a conservative estimate considering the DY scaling factors computed via TFractionFitter method and

presented in Table 5.2. The contributions from backgrounds originating from all other processes are estimated from the simulation and the combination of these sources is conservatively varied up and down by $\pm 30\%$.

- **Decay Branching Fractions**
  POWHEGV2 assumes that the $t\bar{t}$ branching fractions are flavor-independent with flavor-averaged branching fractions of 0.1091, so correction factors were calculated and applied to correct each dilepton channel to the flavor-specific PDG value. The following flavor-specific branching fractions are taken from PDG [6]: $b_{W\to e\nu} = 0.1071 \pm 0.0016$, $b_{W\to\mu\nu} = 0.1063 \pm 0.0015$, and $b_{W\to\tau\nu} = 0.1138 \pm 0.0021$. The branching fraction of τ leptons decaying into a muon is given by $b_{\tau\to\mu\nu\nu} = 0.1741 \pm 0.0004$ and for taus decaying to an electron it is $b_{\tau\to e\nu\nu} = 0.1783 \pm 0.0004$. The following $t\bar{t}$ dilepton branching fraction global correction factors were applied per prompt dilepton channel: $C_{e^+e^-} = 0.9643$, $C_{e^\pm\mu^\mp} = 0.9571$, and $C_{\mu^+\mu^-} = 0.9499$. Dileptonic $t\bar{t}$ events with at least one leptonic tau decay are considered as signal in this analysis and the following branching fraction global correction factors were applied per dilepton channel via tau: $C_{e^+e^-} = 1.0298$, $C_{e^\pm\mu^\mp} = 1.0262$, and $C_{\mu^+\mu^-} = 1.0227$. The measurement is repeated with the branching fractions varied up and down by 1.5%.

6.7.2 Model Systematics

Model systematic uncertainties arise from assumptions such as parameters used to simulate MC events and can impact expected signal and background rates and distributions. The impact of modeling assumptions is determined by repeating the analysis with dedicated simulation samples with altered parameters or by reweighting simulated events.

- **Matrix Element Scales**
  The uncertainty on the ME calculation of the hard production process is assessed by varying the renormalization and factorization scales (μ_R and μ_F) in the POWHEGV2 sample via the use of event weights. Three separate variations are computed varying μ_R and μ_F with respect to their nominal values:

  - μ_R fixed, μ_F varied by 2.0 (0.5) for up (down) variation,
  - μ_F fixed, μ_R varied by 2.0 (0.5) for up (down) variation,
  - μ_R and μ_F varied simultaneously by 2.0 (0.5) for up (down) variation,

  The three ME variations described above are enveloped to obtain the total up and down scale variations.
- **Parton Distribution Functions**
  The uncertainty arising from the PDFs is assessed by reweighting the $t\bar{t}$ signal sample by the 101 weights in the NNPDF3.1 set of PDF errors. Also, a separate variation of the α_S value, used in the NNPDF3.1 PDF set, is performed according to its uncertainties and is propagated through the analysis chain via event weights.

- **Variation of α_S in Parton Shower**
Event weights are used to evaluate the impact of the choice of the α_S value in the PYTHIA8 parton shower simulation. The uncertainties due to the ISR and FSR are estimated via separate up and down variations of the α_S^{ISR} and α_S^{FSR}. Both α_S^{ISR} and α_S^{FSR} are treated as uncorrelated sources of uncertainty.
- **Matrix Element and Parton Shower Matching**
The uncertainty for the matching of the matrix element emissions to those of the parton shower is evaluated with dedicated POWHEGV2+PYTHIA8 MC simulation samples with up and down variations of the $h_{damp} = 1.379^{+0.926}_{-0.505} m_t$ parameter with respect to its nominal value [7].
- **Underlying Event Tune**
An underlying event tune is a set of parameters of the MC event generator parton showering algorithm, describing beam-beam remnants (BBR), particles arising from multiple parton interactions (MPI), hadronization, initial-state radiation (ISR), final-state radiation (FSR), etc., that has been adjusted to better describe some aspects of the data. The choice of the underlying event tune, which is CP5 for all simulated samples used in this analysis, can impact the results of the measurements, and the related uncertainty is estimated using dedicated POWHEGV2+PYTHIA8 simulations with up/down variations of the CP5 tune parameters accordingly to their uncertainties [7].
- **Color Reconnection**
The CR model used in the nominal $t\bar{t}$ simulation is an MPI-based scheme with the early resonance decays (ERD) switched off, as implemented in PYTHIA8. The measurement is repeated using dedicated $t\bar{t}$ samples corresponding to three alternative CR schemes: an MPI-based scheme with ERD switched on, a QCD-inspired scheme, and a gluon-move scheme. The total systematic uncertainty due to CR effects is an envelope of uncertainties from the three alternative CR schemes about the nominal results using the default model.
- **Semi-Leptonic Branching Fraction**
A reweighting of $t\bar{t}$ signal events, with weights obtained via value maps in the EDM analyzer applied to gen jets, is done to correct the generator-level PYTHIA8 semi-leptonic B decay tables to the measured PDG branching ratios and their corresponding uncertainties.
- **Top Quark Mass**
To estimate the impact on the analysis of the uncertainty related to the mass of the top quark, which is assumed to be 172.5 GeV in the default simulation, the measurement is repeated using two dedicated samples assuming top quark masses of 173.5 and 171.5 GeV.
- **Top Quark Transverse Momentum**
The p_T spectrum of top quarks in $t\bar{t}$ data has been found to be softer than modeled by MC simulations. To assess the systematic variation of correcting the top quark p_T in the MC simulation to that measured in data, we use the top p_T reweighting function: $e^{p+q \cdot p_T}$, with $p = 0.0615$ and $q = -0.0005$, derived from data recorded at $\sqrt{s} = 13$ TeV. For each event, the overall weight is the

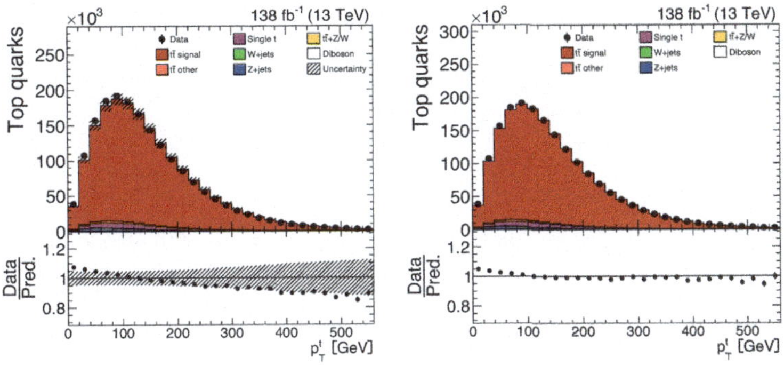

Fig. 6.2 Distribution of top p_T for Full Run II in the combined channel before (left) and after (right) top p_T reweighting

geometric average of the weights obtained for the top quark and the top anti-quark. Figure 6.2 demonstrates the effect of reweighting function on the top p_T spectrum.

References

1. CMS Collaboration, Measurement of the top quark polarization and tt̄ spin correlations using dilepton final states in proton-proton collisions at $\sqrt{s} = 13$ TeV. Phys. Rev. D **100**(7), 072002 (2019). Preliminary results. https://doi.org/10.1103/PhysRevD.100.072002. arXiv: 1907.03729. [Online]. Available: https://cds.cern.ch/record/2681777
2. S. Schmitt, TUnfold, an algorithm for correcting migration effects in high energy physics. JINST **7**, T10003 (2012). https://doi.org/10.1088/1748-0221/7/10/T10003. arXiv: 1205.6201 [physics.data-an]
3. CMS Collaboration, Precision luminosity measurement in proton-proton collisions at $\sqrt{s} = 13$ TeV in 2015 and 2016 at CMS. Eur. Phys. J. C **81**(9), 800 (2021)
4. CMS Collaboration, CMS luminosity measurement for the 2017 data-taking period at $\sqrt{s} = 13$ TeV. CMS Physics Analysis Summary CMS-PAS-LUM-17-004 (2018). [Online]. Available: http://cds.cern.ch/record/2621960
5. CMS Collaboration, CMS luminosity measurement for the 2018 data-taking period at $\sqrt{s} = 13$ TeV. CMS Physics Analysis Summary CMS-PAS-LUM-18-002 (2019). [Online]. Available: http://cds.cern.ch/record/2676164
6. Particle Data Group, Review of particle physics. PTEP **2022**, 083C01 (2022). https://doi.org/10.1093/ptep/ptac097
7. CMS Collaboration, Extraction and validation of a new set of CMS PYTHIA8 tunes from underlying-event measurements. Eur. Phys. J. C **80**(1), 4 (2020). All the figures and tables can be found at http://cms-results.web.cern.ch/cms-results/public-results/publications/GEN-17-001 (CMS Public Pages). https://doi.org/10.1140/epjc/s10052-019-7499-4. arXiv: 1903.12179. [Online]. Available: https://cds.cern.ch/record/2669320

Chapter 7
Results

Abstract This chapter presents the main results of this dissertation—the measured top quark polarization and spin correlation observables, as well as the spin density coefficients that have been extracted from the differential cross-sections. A detailed analysis of the observed distributions and their implications for SM and BSM physics is included. The results are compared with theoretical predictions and their precision relative to previous measurements is noted to emphasize the significance of the findings as a contribution to the field of particle physics.

At the time of writing this dissertation, any results shown in this section are not explicitly approved for public release by CMS and are to be regarded as private, in-progress work utilizing CMS recorded data and MC simulations. By the time the embargo on this dissertation has been lifted, the results shown in this section will have been superseded by CMS public results. Thus, the reader is advised to refer to the CMS public results instead.'

7.1 Measured Top Quark Polarization and $t\bar{t}$ Spin Correlation Observables and Cross-Sections

The set[1] of one-dimensional reconstructed detector-level distributions, absolute differential cross-sections unfolded to parton-level, normalized differential cross-sections unfolded to parton-level, detector response matrices, statistical covariance

[1] A summary of all spin density observables, their respective coefficients, and which coefficient functions of the spin density matrix they probe is provided in Table 6.1.

J. R. Thieman, *Top Quark Spin Correlations with the CMS Detector*,
Springer Theses, https://doi.org/10.1007/978-3-031-76117-1_7

matrices, total systematic covariance matrices, breakdown of uncertainties, and distributions for purity, stability, and efficiency$\times$acceptance is shown for all spin density observables in Figs. 7.1, 7.2, 7.3, 7.4, 7.5, 7.6, 7.7, 7.8, 7.9, 7.10, 7.11, 7.12, 7.13, 7.14, 7.15, 7.16, 7.17, 7.18, 7.19, 7.20, 7.21, 7.22 7.23, 7.24, 7.25, 7.26, 7.27, 7.28, 7.29, 7.30, 7.31, and 7.32, and the two-dimensional set, as a function of $m_{t\bar{t}}$, is shown in Figs. 7.33, 7.34, 7.35, 7.36, 7.37, 7.38, 7.39, 7.40, 7.41, 7.42, 7.43, 7.44, 7.45, 7.46, 7.47, 7.48, 7.49, 7.50, 7.51, 7.52, 7.53 7.54, 7.55, 7.56, 7.57, 7.58 7.59, 7.60, 7.61, 7.62, 7.63, and 7.64.

The presence of the spin correlations is most evident in the slope of the unfolded distribution of $\cos\varphi$, which would be flat in the case of their absence. In agreement with predictions, sizeable asymmetries are also observed in the C, CP, and T_N-even distributions of $\cos\theta_1^k \cos\theta_2^k$, $\cos\theta_1^r \cos\theta_2^r$, $\cos\theta_1^n \cos\theta_2^n$, and $\cos\theta_1^r \cos\theta_2^k + \cos\theta_1^k \cos\theta_2^r$, while asymmetries the remaining C, CP, and T_N-odd distributions are suppressed (refer to Table 3.2).

The statistical and total systematic covariances exhibit a typical pattern of correlation and anti-correlation arising from unfolding, with systematic correlations being significantly stronger in general, reflecting the relative changes in shape of the different distributions in response to the systematic variations. Overall, the most dominant source of systematic uncertainties are due to JES, electron identification, background normalization, and top p_T modelling.

The measured distributions are compared with predictions from the POWHEGV2+PYTHIA8 MC simulation. There is generally decent agreement between the measured distributions and the predictions, although improved agreement is expected when compared to perturbative calculations for $t\bar{t}$ production at NLO in QCD with electroweak corrections. Due to significantly increased luminosity and several optimizations that reduced background contributions and systematic uncertainties, the measurement precision for one-dimensional normalized unfolded cross-sections is improved by around a factor of two compared to CMS TOP-18-006 [1].

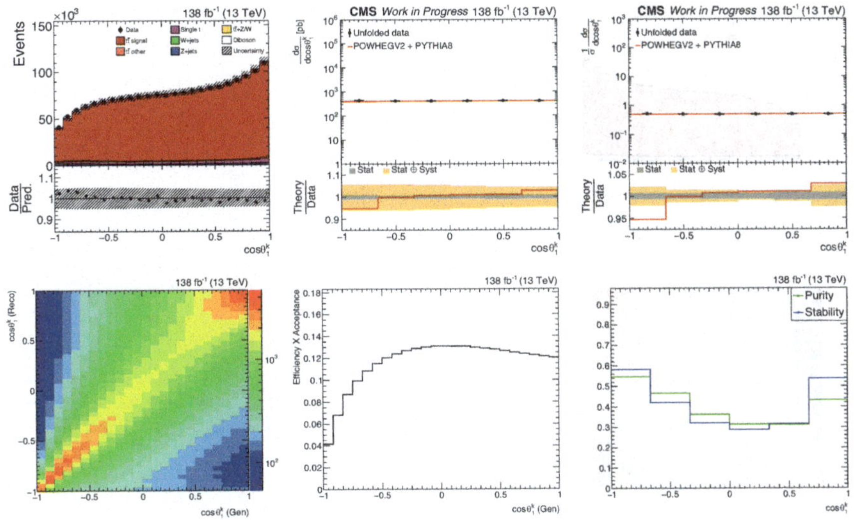

Fig. 7.1 Reconstructed detector-level distribution (Top Left), absolute cross-section unfolded to parton-level (Top Middle), normalized cross-section unfolded to parton-level (Top Right), detector response-matrix (Bottom Left), efficiency×acceptance (Bottom Middle), Purity and Stability (Bottom Right) for polarization observable $\cos\theta_1^k$, from which spin-density coefficient B_1^k (sensitive to spin-density coefficient function b_k^+) is extracted

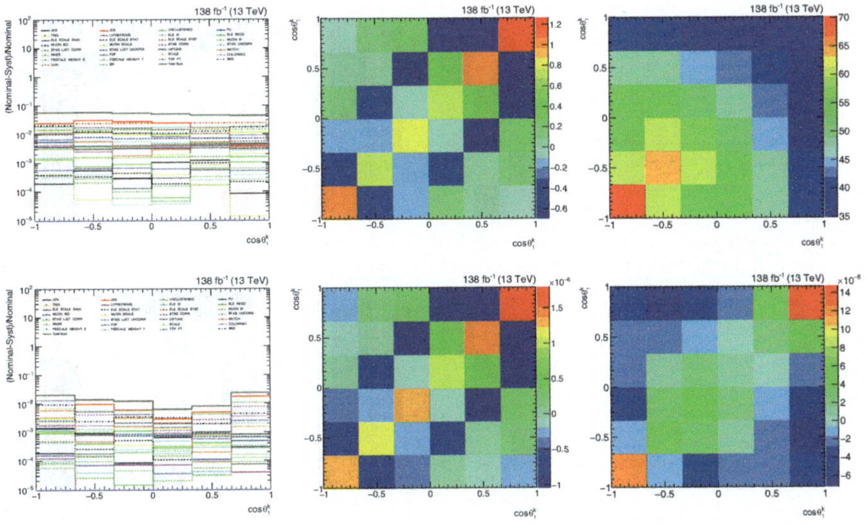

Fig. 7.2 Unfolded-cross section breakdown of uncertainties (Top Left), unfolded cross-section statistical covariance matrix (Top Middle), unfolded cross-section total systematic covariance matrix (Top Right), normalized unfolded-cross section breakdown of uncertainties (Bottom Left), normalized unfolded cross-section statistical covariance matrix (Bottom Middle), normalized unfolded cross-section total systematic covariance matrix (Bottom Right) for polarization observable $\cos\theta_1^k$

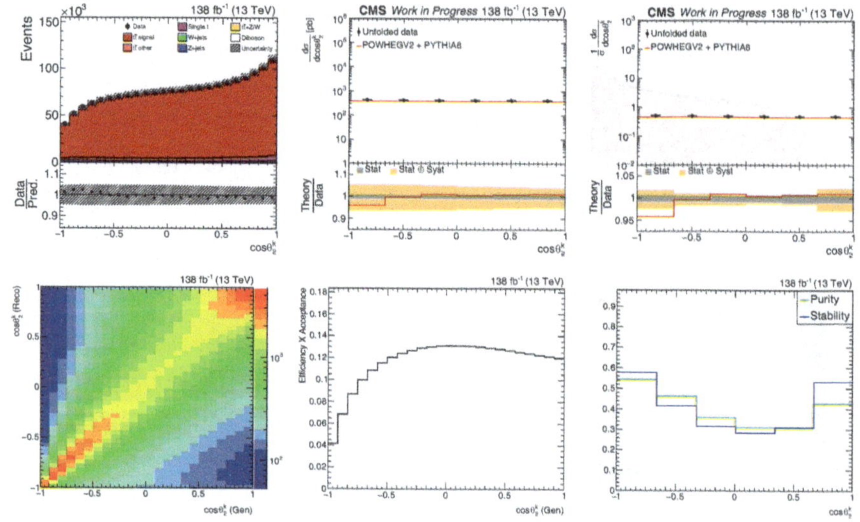

Fig. 7.3 Reconstructed detector-level distribution (Top Left), absolute cross-section unfolded to parton-level (Top Middle), normalized cross-section unfolded to parton-level (Top Right), detector response-matrix (Bottom Left), efficiency×acceptance (Bottom Middle), Purity and Stability (Bottom Right) for polarization observable $\cos\theta_2^k$, from which spin-density coefficient B_2^k (sensitive to spin-density coefficient function b_k^-) is extracted

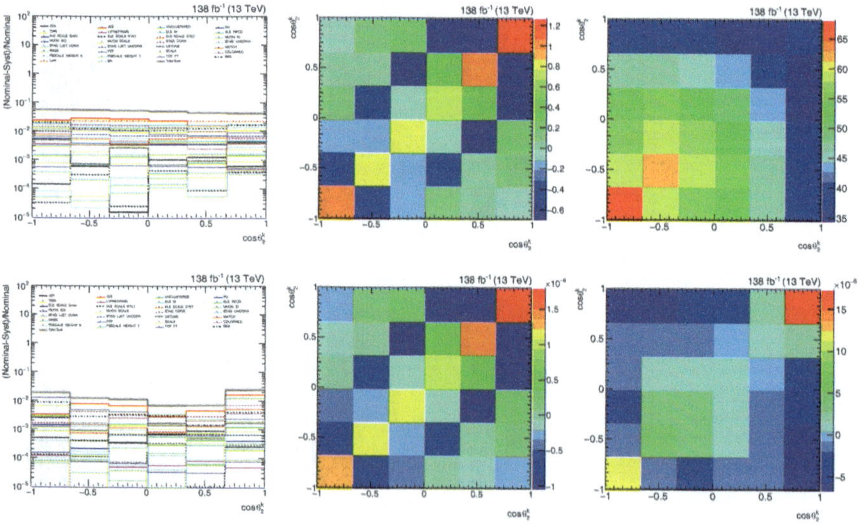

Fig. 7.4 Unfolded-cross section breakdown of uncertainties (Top Left), unfolded cross-section statistical covariance matrix (Top Middle), unfolded cross-section total systematic covariance matrix (Top Right), normalized unfolded-cross section breakdown of uncertainties (Bottom Left), normalized unfolded cross-section statistical covariance matrix (Bottom Middle), normalized unfolded cross-section total systematic covariance matrix (Bottom Right) for polarization observable $\cos\theta_2^k$

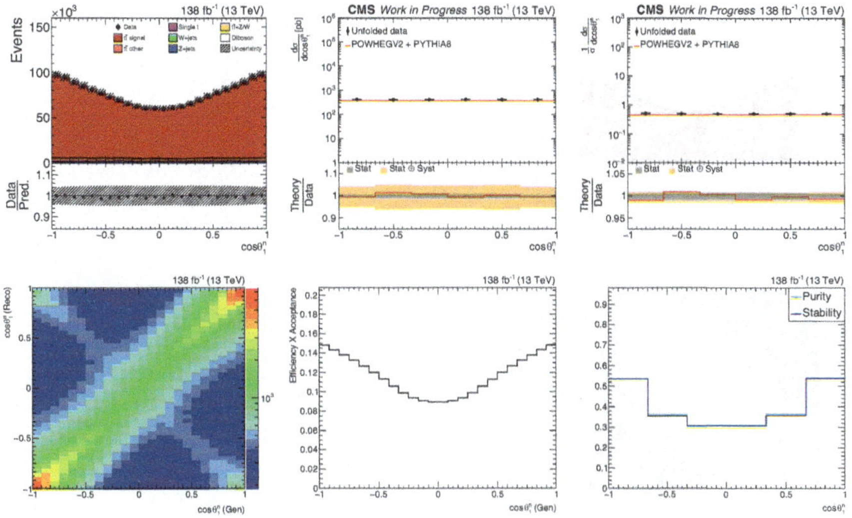

Fig. 7.5 Reconstructed detector-level distribution (Top Left), absolute cross-section unfolded to parton-level (Top Middle), normalized cross-section unfolded to parton-level (Top Right), detector response-matrix (Bottom Left), efficiency×acceptance (Bottom Middle), Purity and Stability (Bottom Right) for polarization observable $\cos\theta_1^n$, from which spin-density coefficient B_1^n (sensitive to spin-density coefficient function b_n^+) is extracted

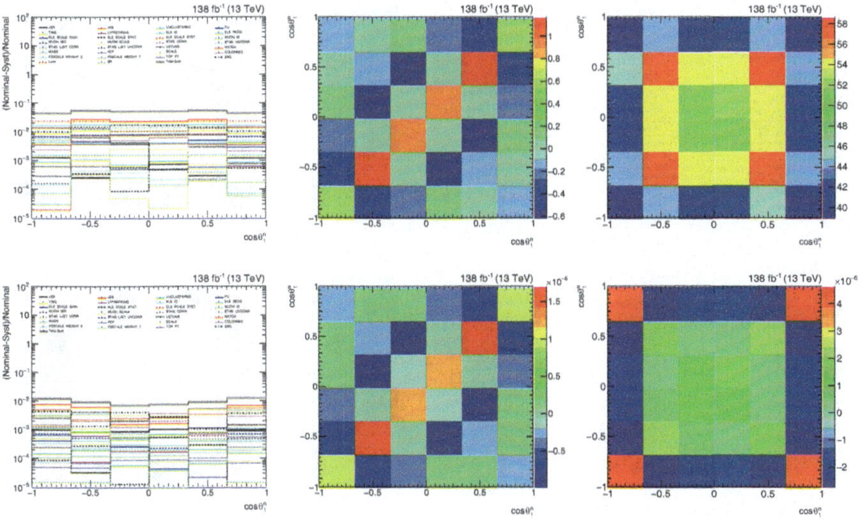

Fig. 7.6 Unfolded-cross section breakdown of uncertainties (Top Left), unfolded cross-section statistical covariance matrix (Top Middle), unfolded cross-section total systematic covariance matrix (Top Right), normalized unfolded-cross section breakdown of uncertainties (Bottom Left), normalized unfolded cross-section statistical covariance matrix (Bottom Middle), normalized unfolded cross-section total systematic covariance matrix (Bottom Right) for polarization observable $\cos\theta_1^n$

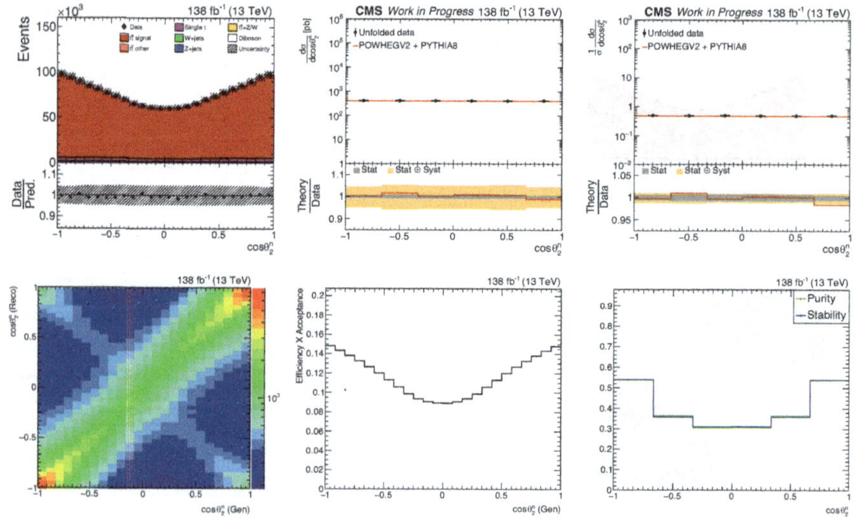

Fig. 7.7 Reconstructed detector-level distribution (Top Left), absolute cross-section unfolded to parton-level (Top Middle), normalized cross-section unfolded to parton-level (Top Right), detector response-matrix (Bottom Left), efficiency×acceptance (Bottom Middle), Purity and Stability (Bottom Right) for polarization observable $\cos\theta_2^n$, from which spin-density coefficient B_2^n (sensitive to spin-density coefficient function b_n^-) is extracted

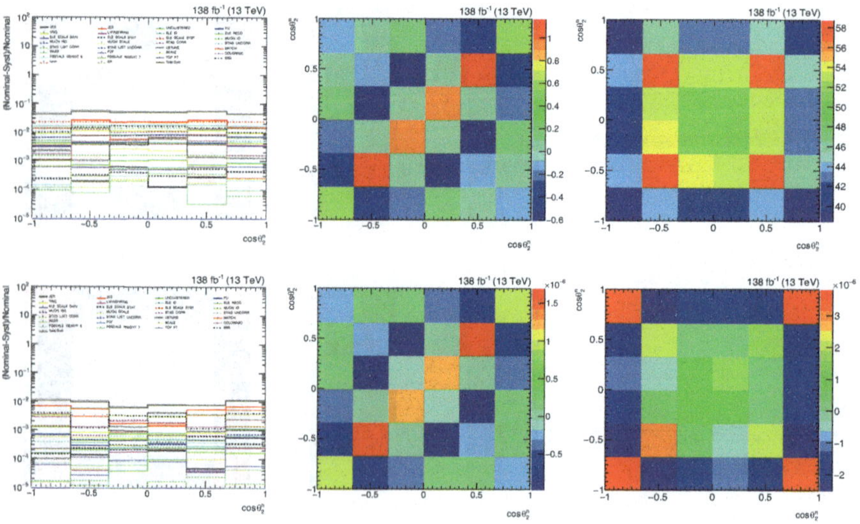

Fig. 7.8 Unfolded-cross section breakdown of uncertainties (Top Left), unfolded cross-section statistical covariance matrix (Top Middle), unfolded cross-section total systematic covariance matrix (Top Right), normalized unfolded-cross section breakdown of uncertainties (Bottom Left), normalized unfolded cross-section statistical covariance matrix (Bottom Middle), normalized unfolded cross-section total systematic covariance matrix (Bottom Right) for polarization observable $\cos\theta_2^n$

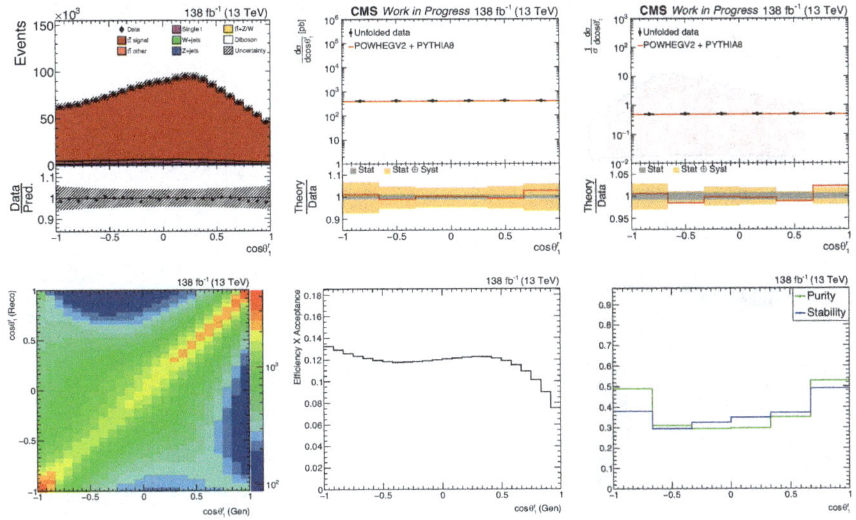

Fig. 7.9 Reconstructed detector-level distribution (Top Left), absolute cross-section unfolded to parton-level (Top Middle), normalized cross-section unfolded to parton-level (Top Right), detector response-matrix (Bottom Left), efficiency×acceptance (Bottom Middle), Purity and Stability (Bottom Right) for polarization observable $\cos\theta_1^r$, from which spin-density coefficient B_1^r (sensitive to spin-density coefficient function b_r^+) is extracted

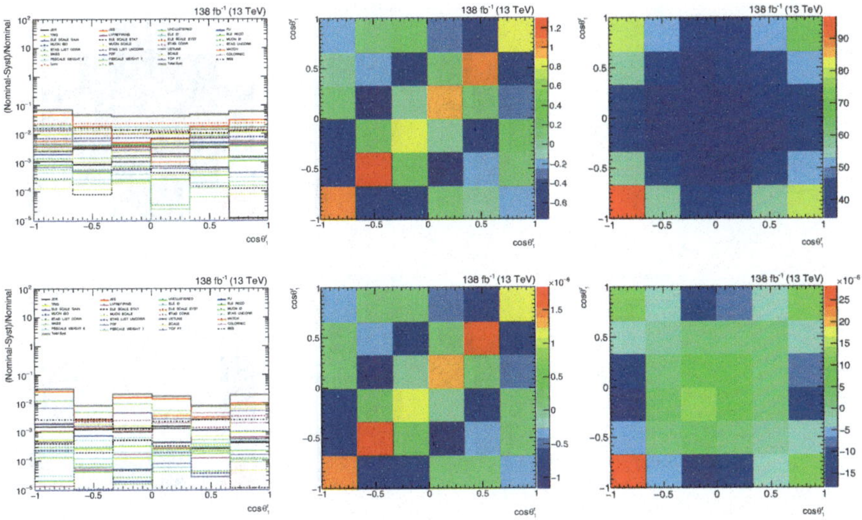

Fig. 7.10 Unfolded-cross section breakdown of uncertainties (Top Left), unfolded cross-section statistical covariance matrix (Top Middle), unfolded cross-section total systematic covariance matrix (Top Right), normalized unfolded-cross section breakdown of uncertainties (Bottom Left), normalized unfolded cross-section statistical covariance matrix (Bottom Middle), normalized unfolded cross-section total systematic covariance matrix (Bottom Right) for polarization observable $\cos\theta_1^r$

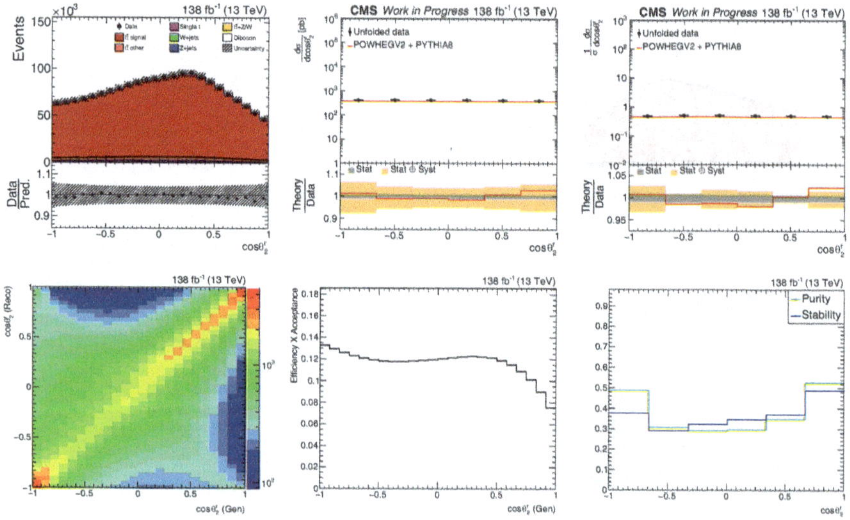

Fig. 7.11 Reconstructed detector-level distribution (Top Left), absolute cross-section unfolded to parton-level (Top Middle), normalized cross-section unfolded to parton-level (Top Right), detector response-matrix (Bottom Left), efficiency×acceptance (Bottom Middle), Purity and Stability (Bottom Right) for polarization observable $\cos\theta_2^r$, from which spin-density coefficient B_2^r (sensitive to spin-density coefficient function b_r^-) is extracted

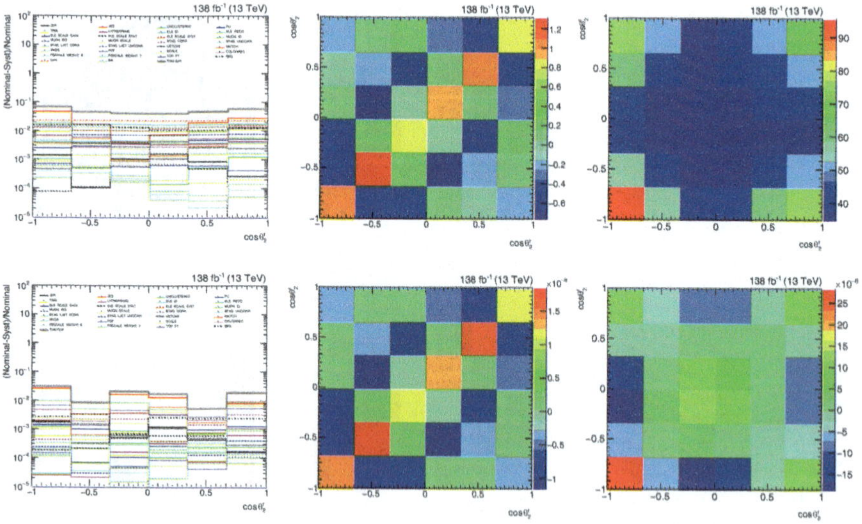

Fig. 7.12 Unfolded-cross section breakdown of uncertainties (Top Left), unfolded cross-section statistical covariance matrix (Top Middle), unfolded cross-section total systematic covariance matrix (Top Right), normalized unfolded-cross section breakdown of uncertainties (Bottom Left), normalized unfolded cross-section statistical covariance matrix (Bottom Middle), normalized unfolded cross-section total systematic covariance matrix (Bottom Right) for polarization observable $\cos\theta_2^r$

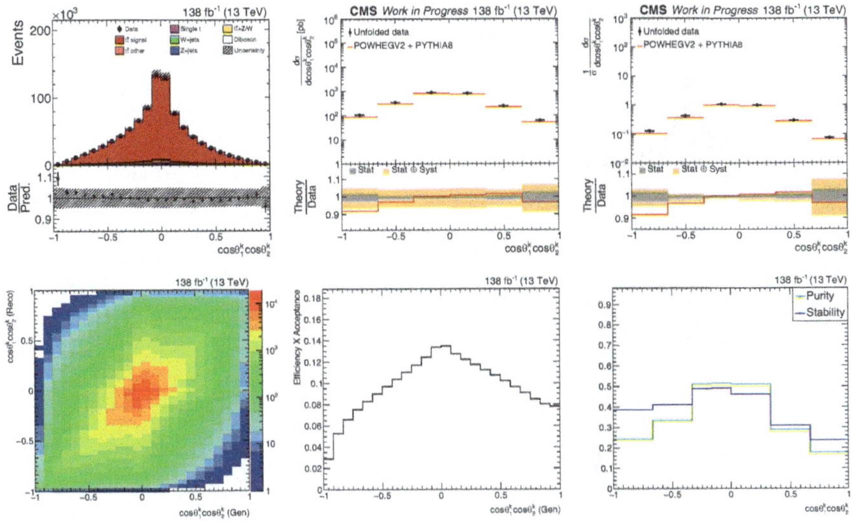

Fig. 7.13 Reconstructed detector-level distribution (Top Left), absolute cross-section unfolded to parton-level (Top Middle), normalized cross-section unfolded to parton-level (Top Right), detector response-matrix (Bottom Left), efficiency×acceptance (Bottom Middle), Purity and Stability (Bottom Right) for diagonal spin correlation observable $\cos\theta_1^k \cos\theta_2^k$, from which spin-density coefficient C_{kk} (sensitive to spin-density coefficient function c_{kk}) is extracted

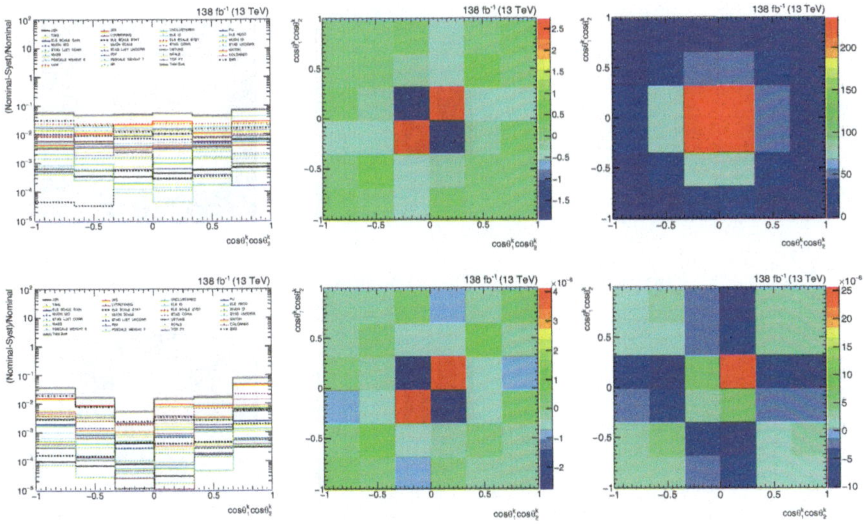

Fig. 7.14 Unfolded-cross section breakdown of uncertainties (Top Left), unfolded cross-section statistical covariance matrix (Top Middle), unfolded cross-section total systematic covariance matrix (Top Right), normalized unfolded-cross section breakdown of uncertainties (Bottom Left), normalized unfolded cross-section statistical covariance matrix (Bottom Middle), normalized unfolded cross-section total systematic covariance matrix (Bottom Right) for diagonal spin correlation observable $\cos\theta_1^k \cos\theta_2^k$

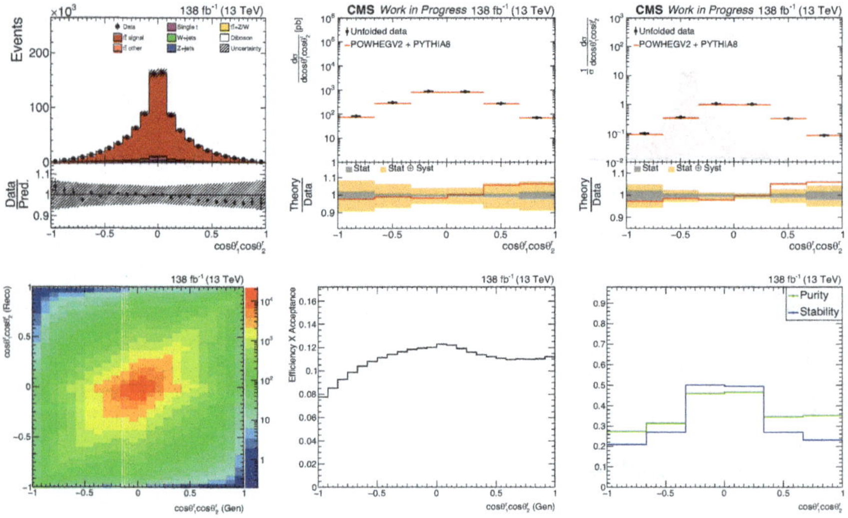

Fig. 7.15 Reconstructed detector-level distribution (Top Left), absolute cross-section unfolded to parton-level (Top Middle), normalized cross-section unfolded to parton-level (Top Right), detector response-matrix (Bottom Left), efficiency×acceptance (Bottom Middle), Purity and Stability (Bottom Right) for diagonal spin correlation observable $\cos\theta_1^r \cos\theta_2^r$, from which spin-density coefficient C_{rr} (sensitive to spin-density coefficient function c_{rr}) is extracted

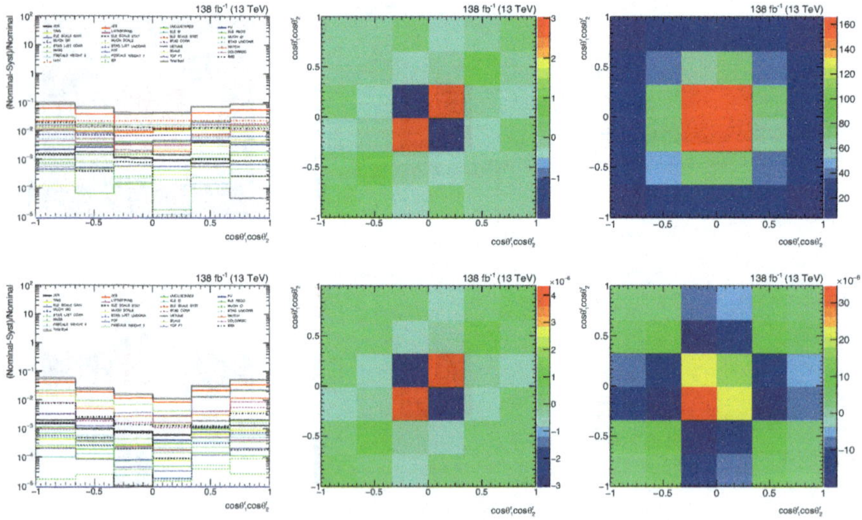

Fig. 7.16 Unfolded-cross section breakdown of uncertainties (Top Left), unfolded cross-section statistical covariance matrix (Top Middle), unfolded cross-section total systematic covariance matrix (Top Right), normalized unfolded-cross section breakdown of uncertainties (Bottom Left), normalized unfolded cross-section statistical covariance matrix (Bottom Middle), normalized unfolded cross-section total systematic covariance matrix (Bottom Right) for diagonal spin correlation observable $\cos\theta_1^r \cos\theta_2^r$

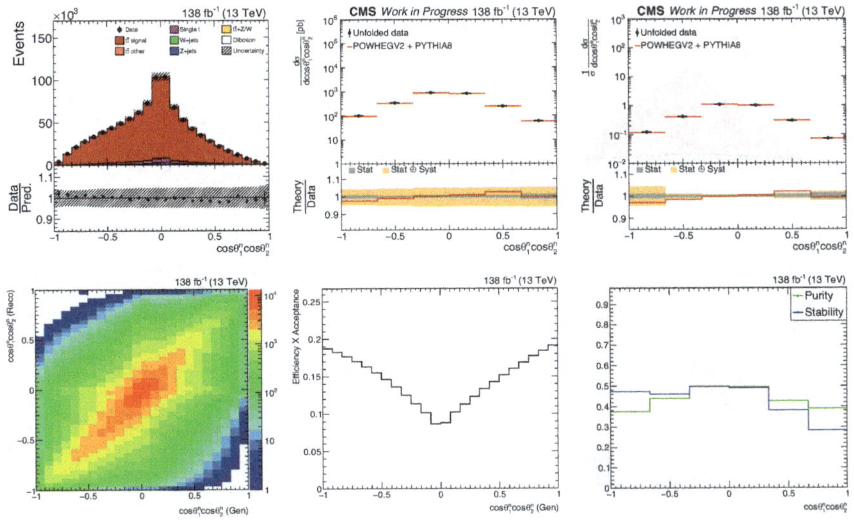

Fig. 7.17 Reconstructed detector-level distribution (Top Left), absolute cross-section unfolded to parton-level (Top Middle), normalized cross-section unfolded to parton-level (Top Right), detector response-matrix (Bottom Left), efficiency×acceptance (Bottom Middle), Purity and Stability (Bottom Right) for diagonal spin correlation observable $\cos\theta_1^n \cos\theta_2^n$, from which spin-density coefficient C_{nn} (sensitive to spin-density coefficient function c_{nn}) is extracted

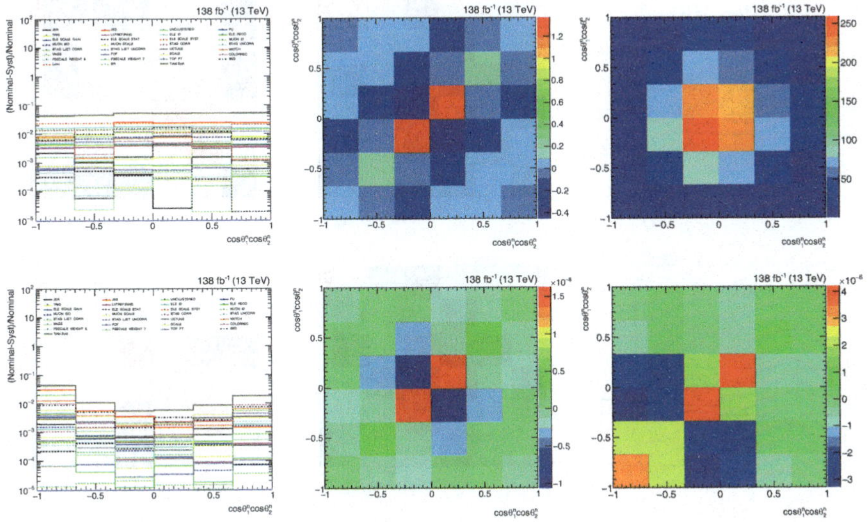

Fig. 7.18 Unfolded-cross section breakdown of uncertainties (Top Left), unfolded cross-section statistical covariance matrix (Top Middle), unfolded cross-section total systematic covariance matrix (Top Right), normalized unfolded-cross section breakdown of uncertainties (Bottom Left), normalized unfolded cross-section statistical covariance matrix (Bottom Middle), normalized unfolded cross-section total systematic covariance matrix (Bottom Right) for diagonal spin correlation observable $\cos\theta_1^n \cos\theta_2^n$

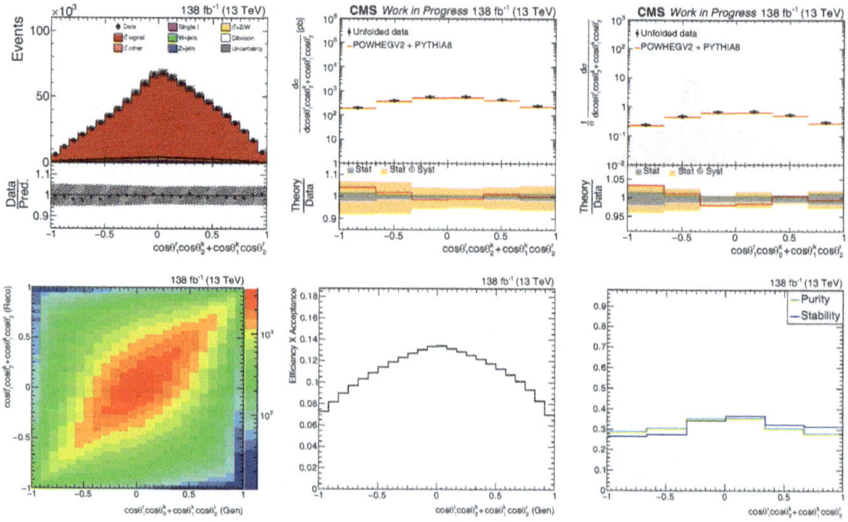

Fig. 7.19 Reconstructed detector-level distribution (Top Left), absolute cross-section unfolded to parton-level (Top Middle), normalized cross-section unfolded to parton-level (Top Right), detector response-matrix (Bottom Left), efficiency×acceptance (Bottom Middle), Purity and Stability (Bottom Right) for off-diagonal spin correlation sum observable $\cos\theta_1^r \cos\theta_2^k + \cos\theta_1^k \cos\theta_2^r$, from which spin-density coefficient $C_{rk} + C_{kr}$ (sensitive to spin-density coefficient function c_{rk}) is extracted

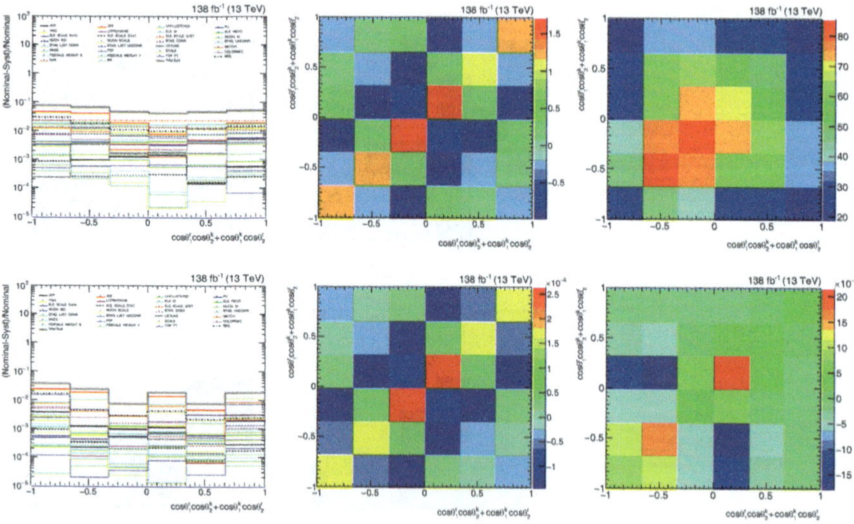

Fig. 7.20 Unfolded-cross section breakdown of uncertainties (Top Left), unfolded cross-section statistical covariance matrix (Top Middle), unfolded cross-section total systematic covariance matrix (Top Right), normalized unfolded-cross section breakdown of uncertainties (Bottom Left), normalized unfolded cross-section statistical covariance matrix (Bottom Middle), normalized unfolded cross-section total systematic covariance matrix (Bottom Right) for off-diagonal spin correlation sum observable $\cos\theta_1^r \cos\theta_2^k + \cos\theta_1^k \cos\theta_2^r$

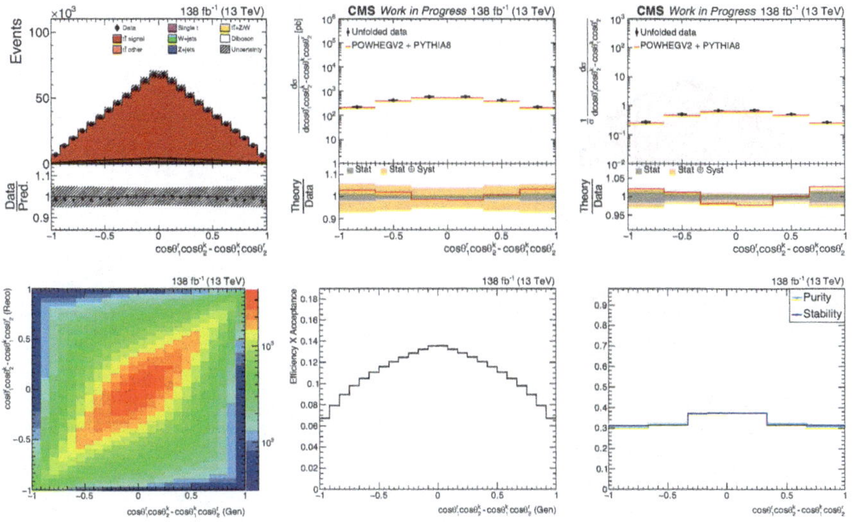

Fig. 7.21 Reconstructed detector-level distribution (Top Left), absolute cross-section unfolded to parton-level (Top Middle), normalized cross-section unfolded to parton-level (Top Right), detector response-matrix (Bottom Left), efficiency×acceptance (Bottom Middle), Purity and Stability (Bottom Right) for off-diagonal spin correlation difference observable $\cos\theta_1^r \cos\theta_2^k - \cos\theta_1^k \cos\theta_2^r$, from which spin-density coefficient $C_{rk} - C_{kr}$ (sensitive to spin-density coefficient function c_n) is extracted

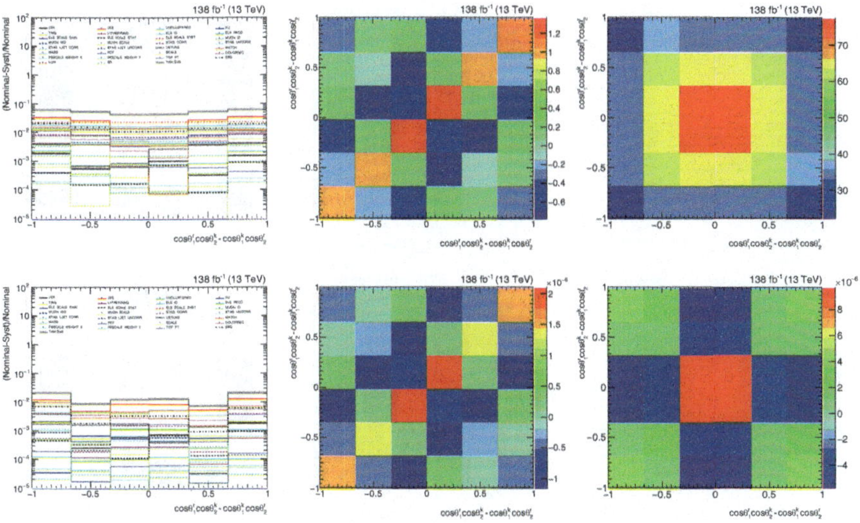

Fig. 7.22 Unfolded-cross section breakdown of uncertainties (Top Left), unfolded cross-section statistical covariance matrix (Top Middle), unfolded cross-section total systematic covariance matrix (Top Right), normalized unfolded-cross section breakdown of uncertainties (Bottom Left), normalized unfolded cross-section statistical covariance matrix (Bottom Middle), normalized unfolded cross-section total systematic covariance matrix (Bottom Right) for off-diagonal spin correlation difference observable $\cos\theta_1^r \cos\theta_2^k - \cos\theta_1^k \cos\theta_2^r$

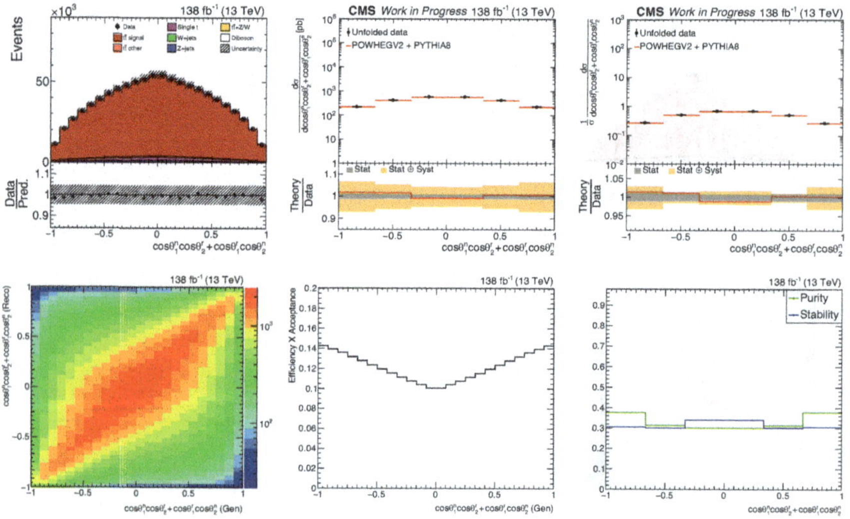

Fig. 7.23 Reconstructed detector-level distribution (Top Left), absolute cross-section unfolded to parton-level (Top Middle), normalized cross-section unfolded to parton-level (Top Right), detector response-matrix (Bottom Left), efficiency×acceptance (Bottom Middle), Purity and Stability (Bottom Right) for off-diagonal spin correlation sum observable $\cos\theta_1^n \cos\theta_2^r + \cos\theta_1^r \cos\theta_2^n$, from which spin-density coefficient $C_{nr} + C_{rn}$ (sensitive to spin-density coefficient function c_{nr}) is extracted

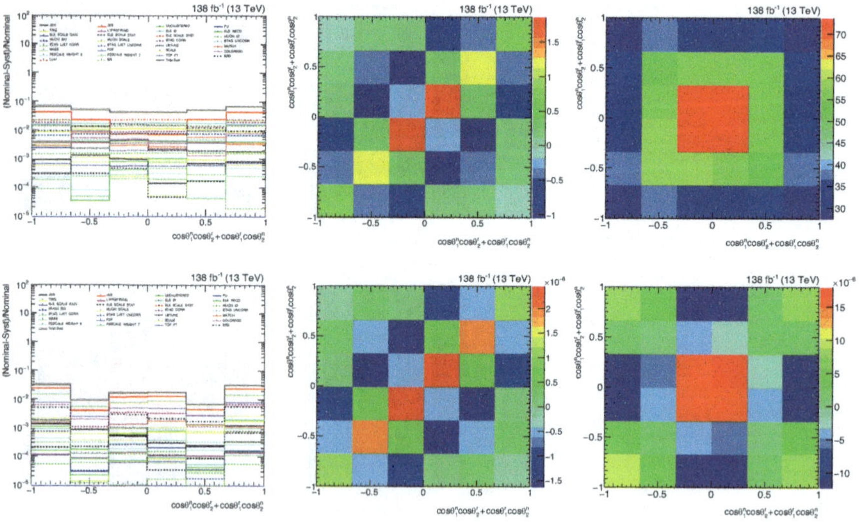

Fig. 7.24 Unfolded-cross section breakdown of uncertainties (Top Left), unfolded cross-section statistical covariance matrix (Top Middle), unfolded cross-section total systematic covariance matrix (Top Right), normalized unfolded-cross section breakdown of uncertainties (Bottom Left), normalized unfolded cross-section statistical covariance matrix (Bottom Middle), normalized unfolded cross-section total systematic covariance matrix (Bottom Right) for off-diagonal spin correlation sum observable $\cos\theta_1^n \cos\theta_2^r + \cos\theta_1^r \cos\theta_2^n$

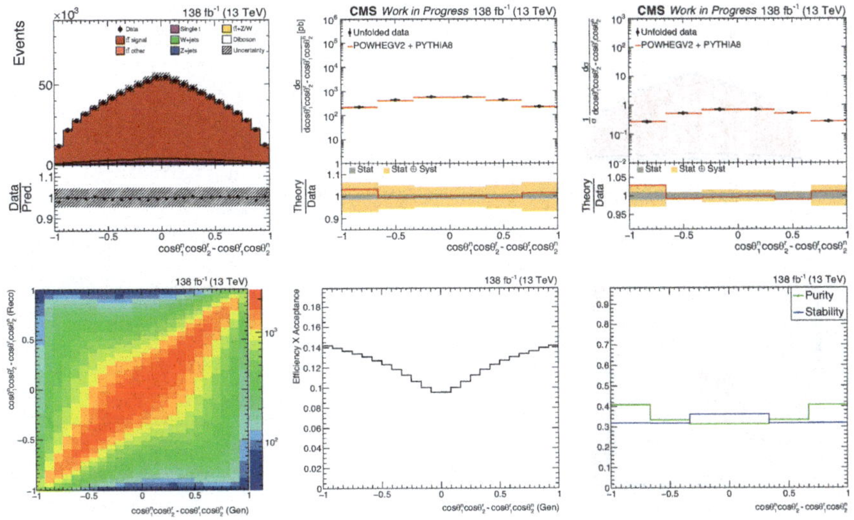

Fig. 7.25 Reconstructed detector-level distribution (Top Left), absolute cross-section unfolded to parton-level (Top Middle), normalized cross-section unfolded to parton-level (Top Right), detector response-matrix (Bottom Left), efficiency×acceptance (Bottom Middle), Purity and Stability (Bottom Right) for off-diagonal spin correlation difference observable $\cos\theta_1^n \cos\theta_2^r - \cos\theta_1^r \cos\theta_2^n$, from which spin-density coefficient $C_{nr} - C_{rn}$ (sensitive to spin-density coefficient function c_k) is extracted

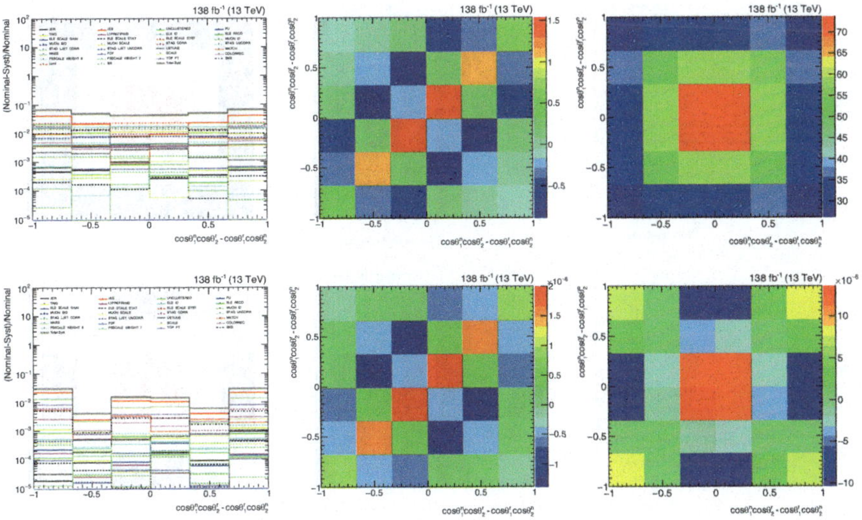

Fig. 7.26 Unfolded-cross section breakdown of uncertainties (Top Left), unfolded cross-section statistical covariance matrix (Top Middle), unfolded cross-section total systematic covariance matrix (Top Right), normalized unfolded-cross section breakdown of uncertainties (Bottom Left), normalized unfolded cross-section statistical covariance matrix (Bottom Middle), normalized unfolded cross-section total systematic covariance matrix (Bottom Right) for off-diagonal spin correlation difference observable $\cos\theta_1^n \cos\theta_2^r - \cos\theta_1^r \cos\theta_2^n$

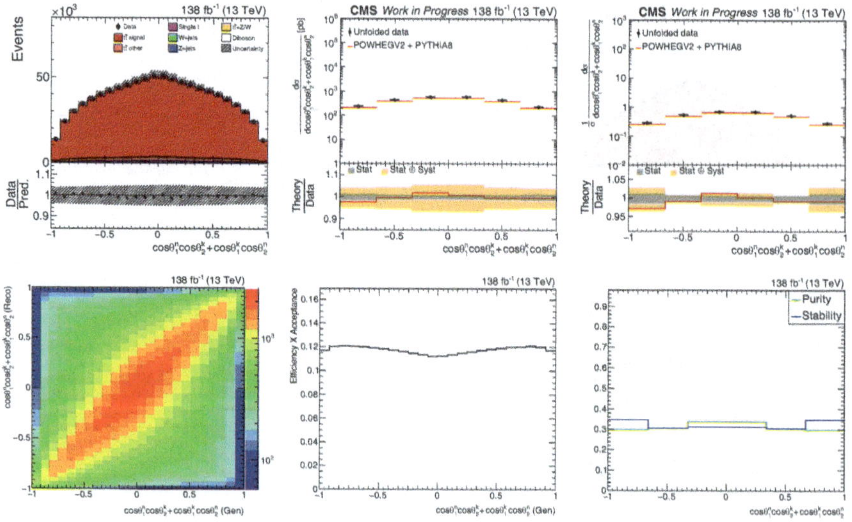

Fig. 7.27 Reconstructed detector-level distribution (Top Left), absolute cross-section unfolded to parton-level (Top Middle), normalized cross-section unfolded to parton-level (Top Right), detector response-matrix (Bottom Left), efficiency×acceptance (Bottom Middle), Purity and Stability (Bottom Right) for off-diagonal spin correlation sum observable $\cos\theta_1^n \cos\theta_2^k + \cos\theta_1^k \cos\theta_2^n$, from which spin-density coefficient $C_{nk} + C_{kn}$ (sensitive to spin-density coefficient function c_{kn}) is extracted

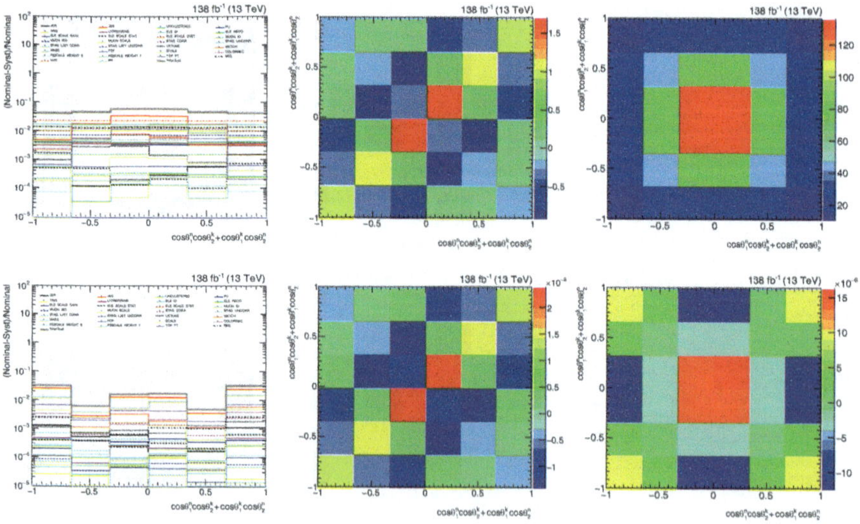

Fig. 7.28 Unfolded-cross section breakdown of uncertainties (Top Left), unfolded cross-section statistical covariance matrix (Top Middle), unfolded cross-section total systematic covariance matrix (Top Right), normalized unfolded-cross section breakdown of uncertainties (Bottom Left), normalized unfolded cross-section statistical covariance matrix (Bottom Middle), normalized unfolded cross-section total systematic covariance matrix (Bottom Right) for off-diagonal spin correlation sum observable $\cos\theta_1^n \cos\theta_2^k + \cos\theta_1^k \cos\theta_2^n$

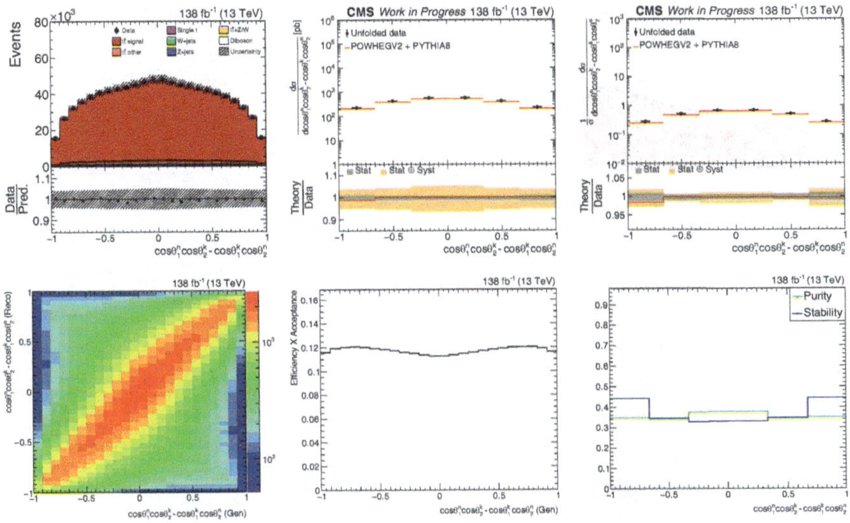

Fig. 7.29 Reconstructed detector-level distribution (Top Left), absolute cross-section unfolded to parton-level (Top Middle), normalized cross-section unfolded to parton-level (Top Right), detector response-matrix (Bottom Left), efficiency×acceptance (Bottom Middle), Purity and Stability (Bottom Right) for off-diagonal spin correlation difference observable $\cos\theta_1^n \cos\theta_2^k - \cos\theta_1^k \cos\theta_2^n$, from which spin-density coefficient $C_{nk} - C_{kn}$ (sensitive to spin-density coefficient function $-c_r$) is extracted

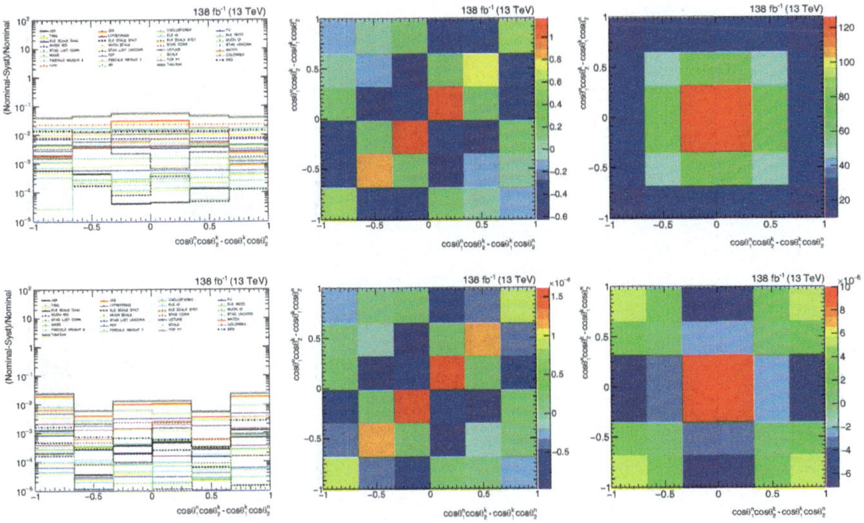

Fig. 7.30 Unfolded-cross section breakdown of uncertainties (Top Left), unfolded cross-section statistical covariance matrix (Top Middle), unfolded cross-section total systematic covariance matrix (Top Right), normalized unfolded-cross section breakdown of uncertainties (Bottom Left), normalized unfolded cross-section statistical covariance matrix (Bottom Middle), normalized unfolded cross-section total systematic covariance matrix (Bottom Right) for off-diagonal spin correlation difference observable $\cos\theta_1^n \cos\theta_2^k - \cos\theta_1^k \cos\theta_2^n$

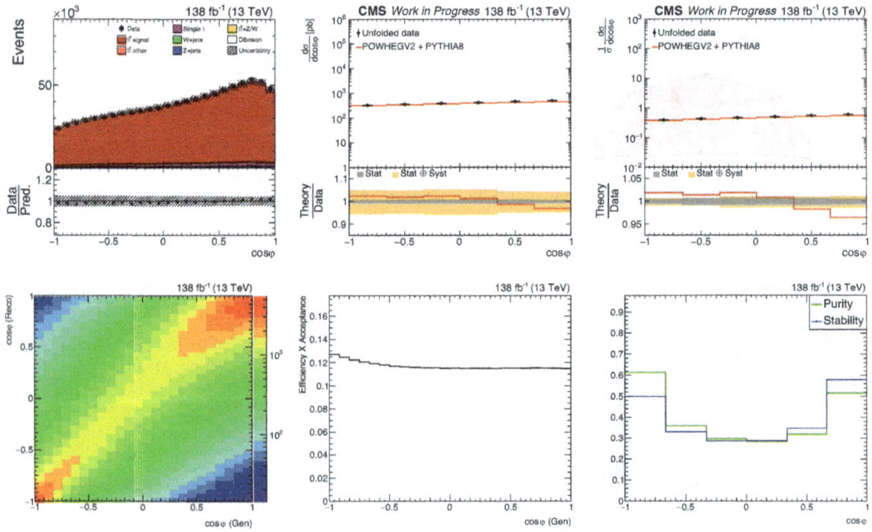

Fig. 7.31 Reconstructed detector-level distribution (Top Left), absolute cross-section unfolded to parton-level (Top Middle), normalized cross-section unfolded to parton-level (Top Right), detector response-matrix (Bottom Left), efficiency×acceptance (Bottom Middle), Purity and Stability (Bottom Right) for observable $\cos\varphi$, from which spin-density coefficient $D = -\frac{1}{3}(C_{kk} + C_{rr} + C_{nn})$ (sensitive to spin-density coefficient function $-\frac{1}{3}(c_{kk} + c_{rr} + c_{nn})$) is extracted

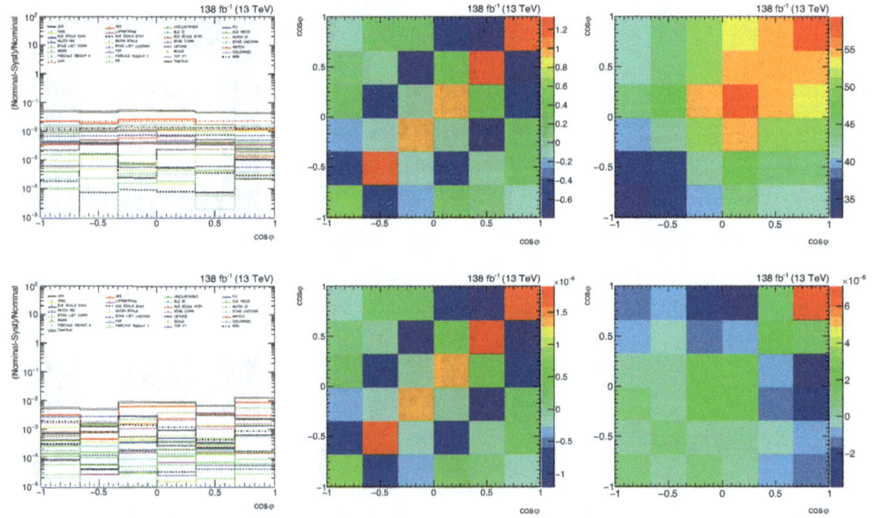

Fig. 7.32 Unfolded-cross section breakdown of uncertainties (Top Left), unfolded cross-section statistical covariance matrix (Top Middle), unfolded cross-section total systematic covariance matrix (Top Right), normalized unfolded-cross section breakdown of uncertainties (Bottom Left), normalized unfolded cross-section statistical covariance matrix (Bottom Middle), normalized unfolded cross-section total systematic covariance matrix (Bottom Right) for observable $\cos\varphi$

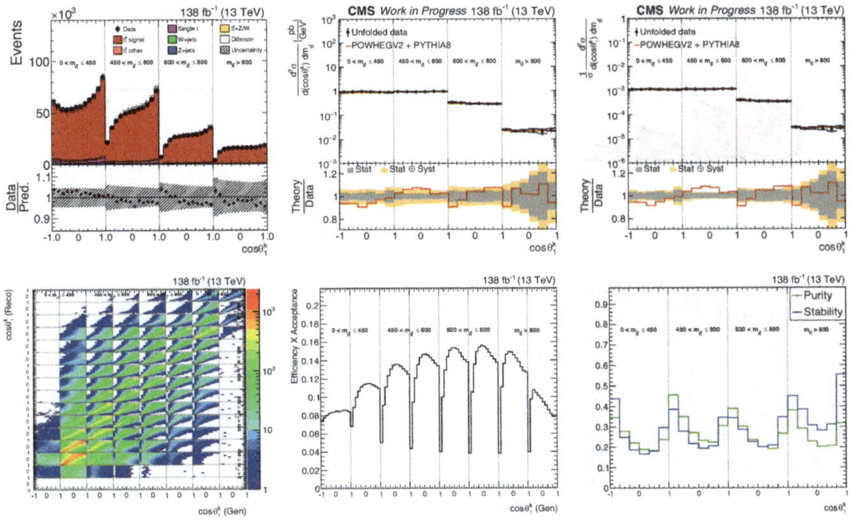

Fig. 7.33 Reconstructed detector-level distribution (Top Left), absolute cross-section unfolded to parton-level (Top Middle), normalized cross-section unfolded to parton-level (Top Right), detector response-matrix (Bottom Left), efficiency×acceptance (Bottom Middle), Purity and Stability (Bottom Right) for polarization observable $\cos\theta_1^k$ as a function of $m_{t\bar{t}}$, from which differential spin-density coefficients B_1^k (sensitive to spin-density coefficient functions b_k^+) are extracted

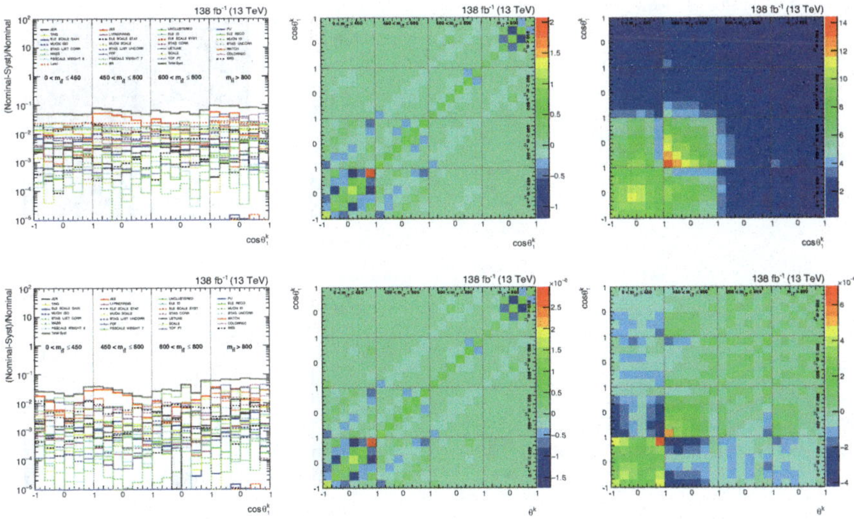

Fig. 7.34 Unfolded-cross section breakdown of uncertainties (Top Left), unfolded cross-section statistical covariance matrix (Top Middle), unfolded cross-section total systematic covariance matrix (Top Right), normalized unfolded-cross section breakdown of uncertainties (Bottom Left), normalized unfolded cross-section statistical covariance matrix (Bottom Middle), normalized unfolded cross-section total systematic covariance matrix (Bottom Right) for polarization observable $\cos\theta_1^k$ as a function of $m_{t\bar{t}}$

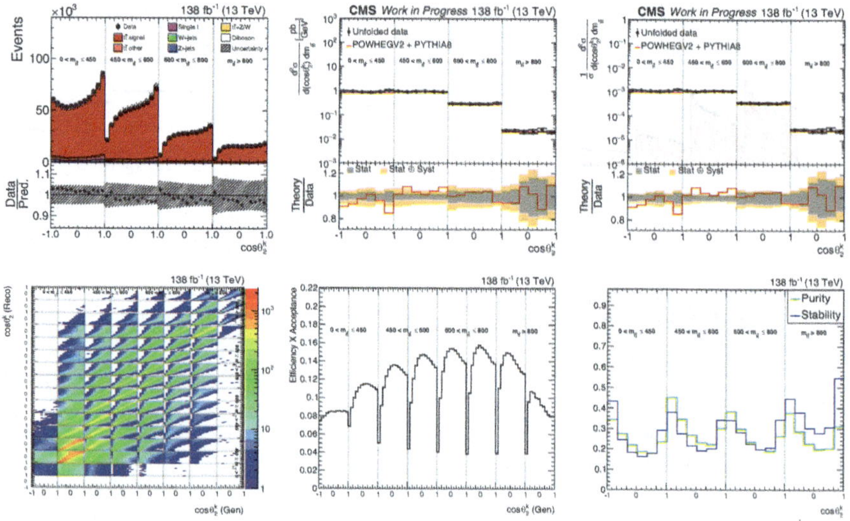

Fig. 7.35 Reconstructed detector-level distribution (Top Left), absolute cross-section unfolded to parton-level (Top Middle), normalized cross-section unfolded to parton-level (Top Right), detector response-matrix (Bottom Left), efficiency×acceptance (Bottom Middle), Purity and Stability (Bottom Right) for polarization observable $\cos\theta_2^k$ as a function of $m_{t\bar{t}}$, from which differential spin-density coefficients B_2^k (sensitive to spin-density coefficient functions b_k^-) are extracted

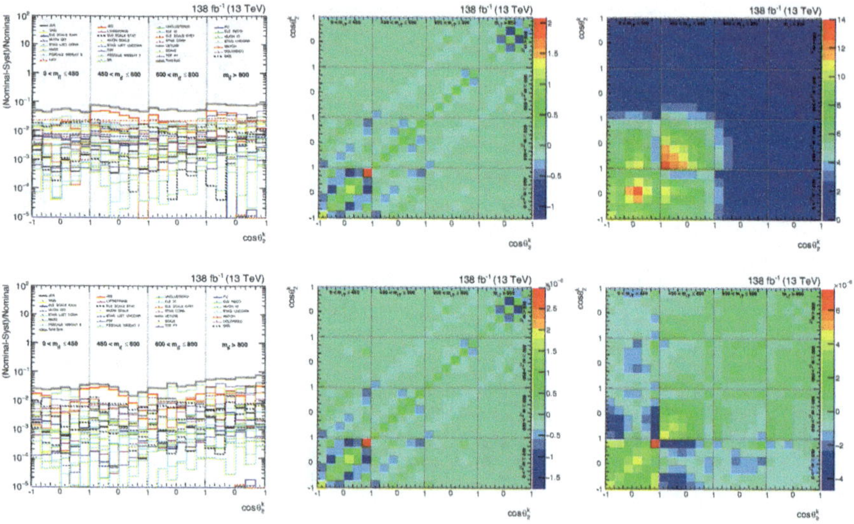

Fig. 7.36 Unfolded-cross section breakdown of uncertainties (Top Left), unfolded cross-section statistical covariance matrix (Top Middle), unfolded cross-section total systematic covariance matrix (Top Right), normalized unfolded-cross section breakdown of uncertainties (Bottom Left), normalized unfolded cross-section statistical covariance matrix (Bottom Middle), normalized unfolded cross-section total systematic covariance matrix (Bottom Right) for polarization observable $\cos\theta_2^k$ as a function of $m_{t\bar{t}}$

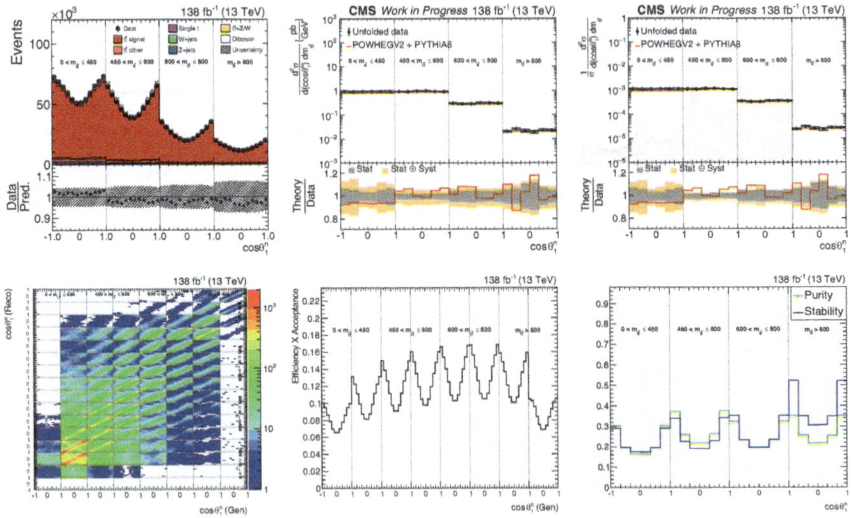

Fig. 7.37 Reconstructed detector-level distribution (Top Left), absolute cross-section unfolded to parton-level (Top Middle), normalized cross-section unfolded to parton-level (Top Right), detector response-matrix (Bottom Left), efficiency×acceptance (Bottom Middle), Purity and Stability (Bottom Right) for polarization observable $\cos\theta_1^n$ as a function of $m_{t\bar{t}}$, from which differential spin-density coefficients B_1^n (sensitive to spin-density coefficient functions b_n^+) are extracted

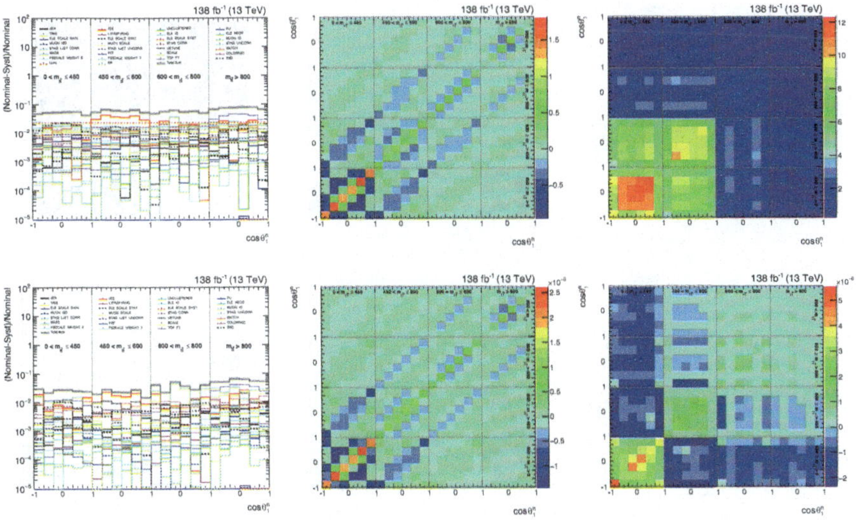

Fig. 7.38 Unfolded-cross section breakdown of uncertainties (Top Left), unfolded cross-section statistical covariance matrix (Top Middle), unfolded cross-section total systematic covariance matrix (Top Right), normalized unfolded-cross section breakdown of uncertainties (Bottom Left), normalized unfolded cross-section statistical covariance matrix (Bottom Middle), normalized unfolded cross-section total systematic covariance matrix (Bottom Right) for polarization observable $\cos\theta_1^n$ as a function of $m_{t\bar{t}}$

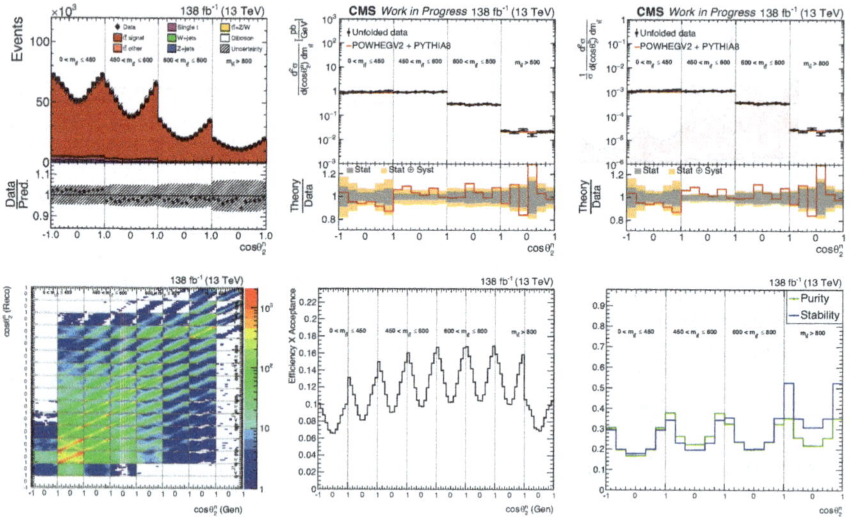

Fig. 7.39 Reconstructed detector-level distribution (Top Left), absolute cross-section unfolded to parton-level (Top Middle), normalized cross-section unfolded to parton-level (Top Right), detector response-matrix (Bottom Left), efficiency×acceptance (Bottom Middle), Purity and Stability (Bottom Right) for polarization observable $\cos\theta_2^n$ as a function of $m_{t\bar{t}}$, from which differential spin-density coefficients B_2^n (sensitive to spin-density coefficient functions b_n^-) are extracted

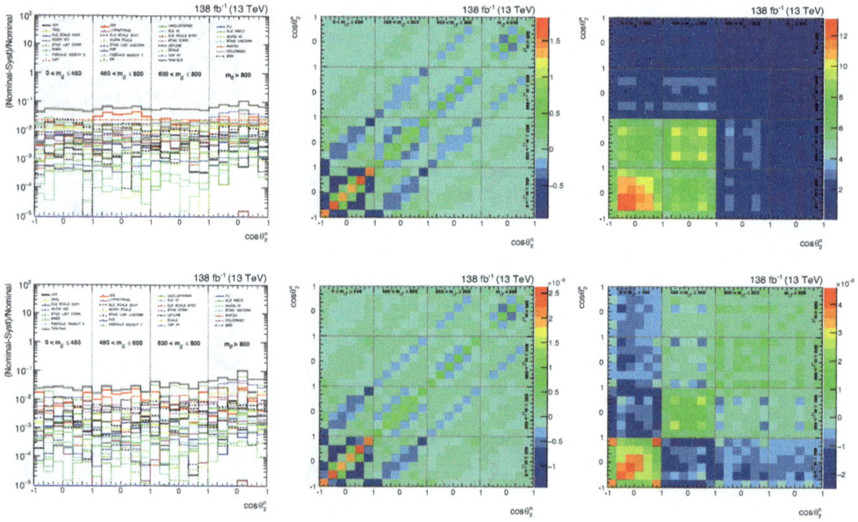

Fig. 7.40 Unfolded-cross section breakdown of uncertainties (Top Left), unfolded cross-section statistical covariance matrix (Top Middle), unfolded cross-section total systematic covariance matrix (Top Right), normalized unfolded-cross section breakdown of uncertainties (Bottom Left), normalized unfolded cross-section statistical covariance matrix (Bottom Middle), normalized unfolded cross-section total systematic covariance matrix (Bottom Right) for polarization observable $\cos\theta_2^n$ as a function of $m_{t\bar{t}}$

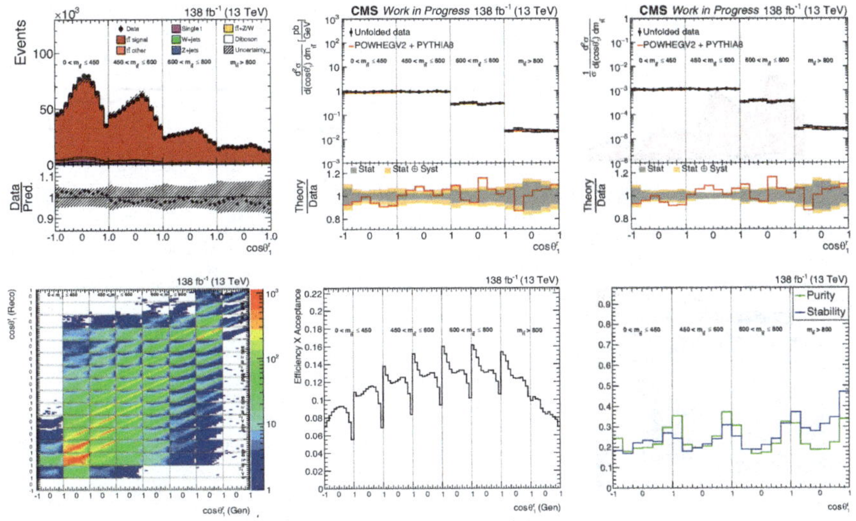

Fig. 7.41 Reconstructed detector-level distribution (Top Left), absolute cross-section unfolded to parton-level (Top Middle), normalized cross-section unfolded to parton-level (Top Right), detector response-matrix (Bottom Left), efficiency×acceptance (Bottom Middle), Purity and Stability (Bottom Right) for polarization observable $\cos\theta_1^r$ as a function of $m_{t\bar{t}}$, from which differential spin-density coefficients B_1^r (sensitive to spin-density coefficient functions b_r^+) are extracted

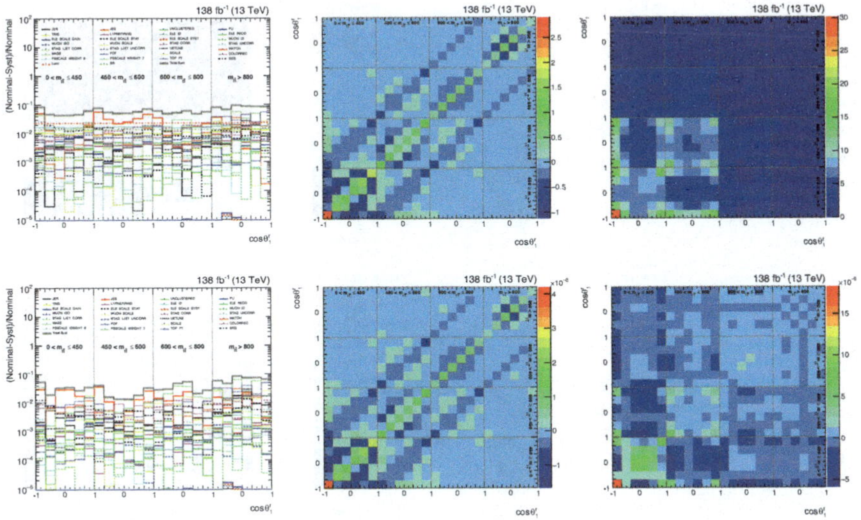

Fig. 7.42 Unfolded-cross section breakdown of uncertainties (Top Left), unfolded cross-section statistical covariance matrix (Top Middle), unfolded cross-section total systematic covariance matrix (Top Right), normalized unfolded-cross section breakdown of uncertainties (Bottom Left), normalized unfolded cross-section statistical covariance matrix (Bottom Middle), normalized unfolded cross-section total systematic covariance matrix (Bottom Right) for polarization observable $\cos\theta_1^r$ as a function of $m_{t\bar{t}}$

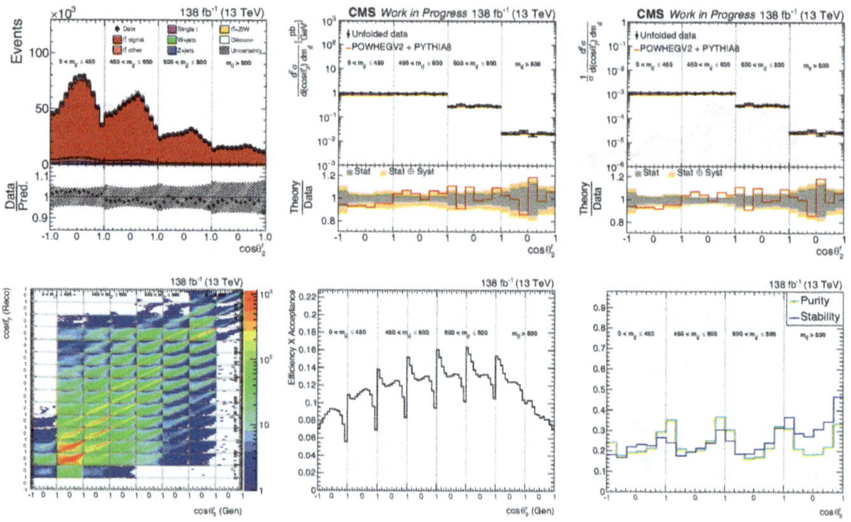

Fig. 7.43 Reconstructed detector-level distribution (Top Left), absolute cross-section unfolded to parton-level (Top Middle), normalized cross-section unfolded to parton-level (Top Right), detector response-matrix (Bottom Left), efficiency×acceptance (Bottom Middle), Purity and Stability (Bottom Right) for polarization observable $\cos\theta_2^r$ as a function of $m_{t\bar{t}}$, from which differential spin-density coefficients B_2^r (sensitive to spin-density coefficient functions b_r^-) are extracted

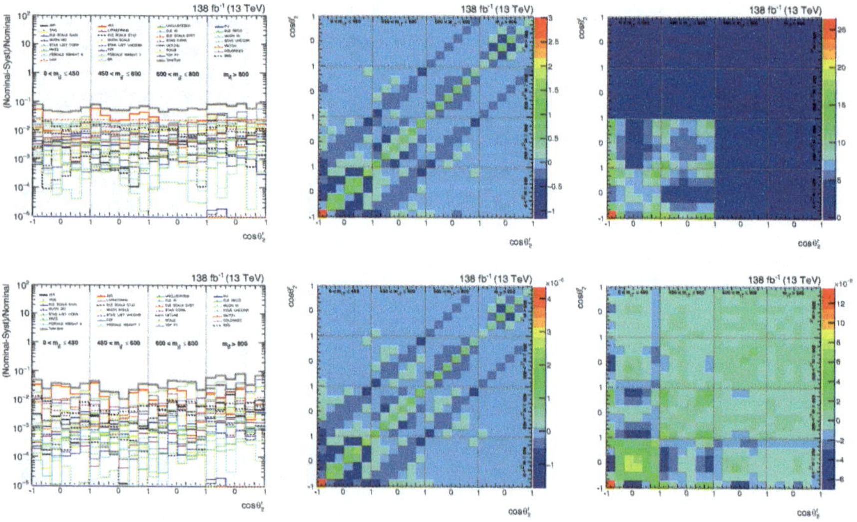

Fig. 7.44 Unfolded-cross section breakdown of uncertainties (Top Left), unfolded cross-section statistical covariance matrix (Top Middle), unfolded cross-section total systematic covariance matrix (Top Right), normalized unfolded-cross section breakdown of uncertainties (Bottom Left), normalized unfolded cross-section statistical covariance matrix (Bottom Middle), normalized unfolded cross-section total systematic covariance matrix (Bottom Right) for polarization observable $\cos\theta_2^r$ as a function of $m_{t\bar{t}}$

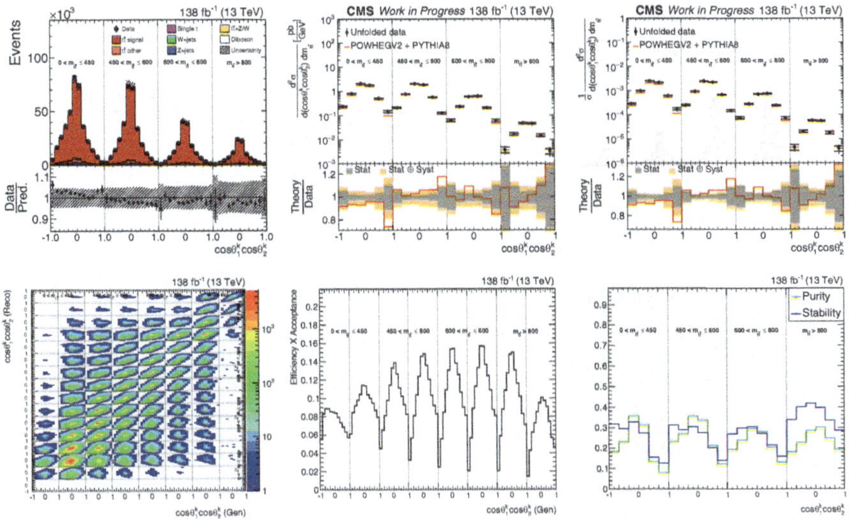

Fig. 7.45 Reconstructed detector-level distribution (Top Left), absolute cross-section unfolded to parton-level (Top Middle), normalized cross-section unfolded to parton-level (Top Right), detector response-matrix (Bottom Left), efficiency×acceptance (Bottom Middle), Purity and Stability (Bottom Right) for diagonal spin correlation observable $\cos\theta_1^k \cos\theta_2^k$ as a function of $m_{t\bar{t}}$, from which differential spin-density coefficients C_{kk} (sensitive to spin-density coefficient functions c_{kk}) are extracted

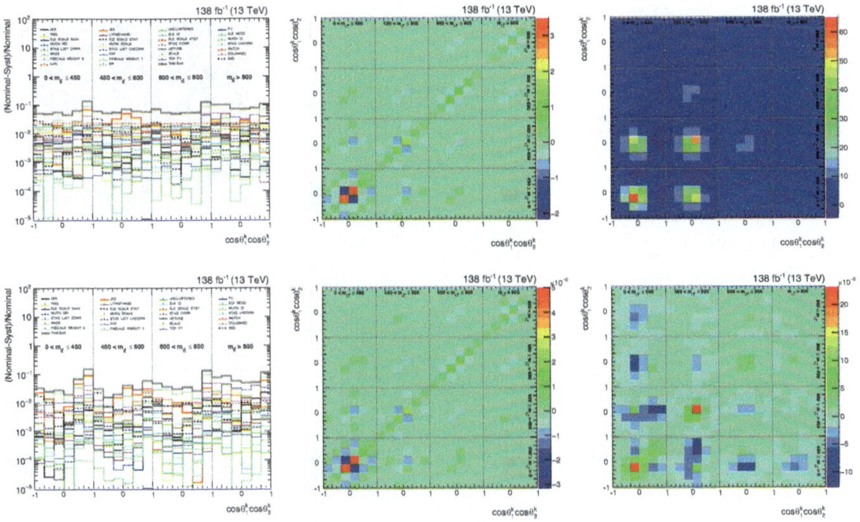

Fig. 7.46 Unfolded-cross section breakdown of uncertainties (Top Left), unfolded cross-section statistical covariance matrix (Top Middle), unfolded cross-section total systematic covariance matrix (Top Right), normalized unfolded-cross section breakdown of uncertainties (Bottom Left), normalized unfolded cross-section statistical covariance matrix (Bottom Middle), normalized unfolded cross-section total systematic covariance matrix (Bottom Right) for diagonal spin correlation observable $\cos\theta_1^k \cos\theta_2^k$ as a function of $m_{t\bar{t}}$

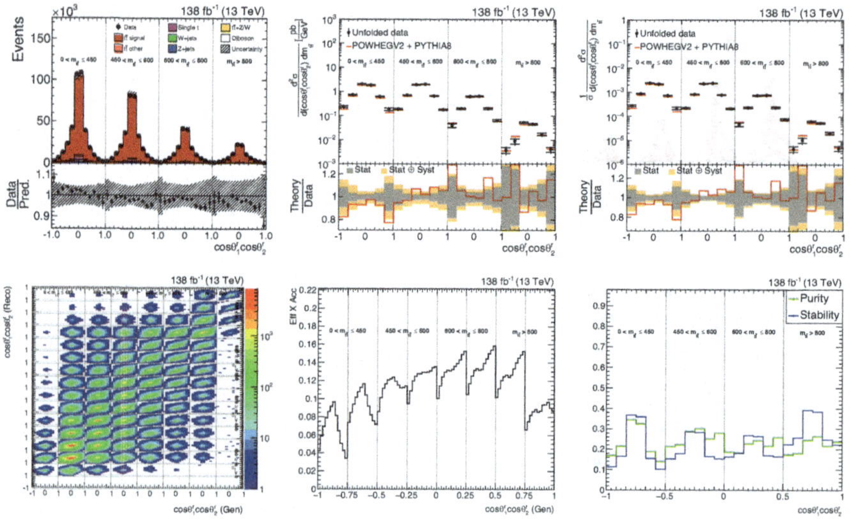

Fig. 7.47 Reconstructed detector-level distribution (Top Left), absolute cross-section unfolded to parton-level (Top Middle), normalized cross-section unfolded to parton-level (Top Right), detector response-matrix (Bottom Left), efficiency×acceptance (Bottom Middle), Purity and Stability (Bottom Right) for diagonal spin correlation observable $\cos\theta_1^r \cos\theta_2^r$ as a function of $m_{t\bar{t}}$, from which differential spin-density coefficients C_{rr} (sensitive to spin-density coefficient functions c_{rr}) are extracted

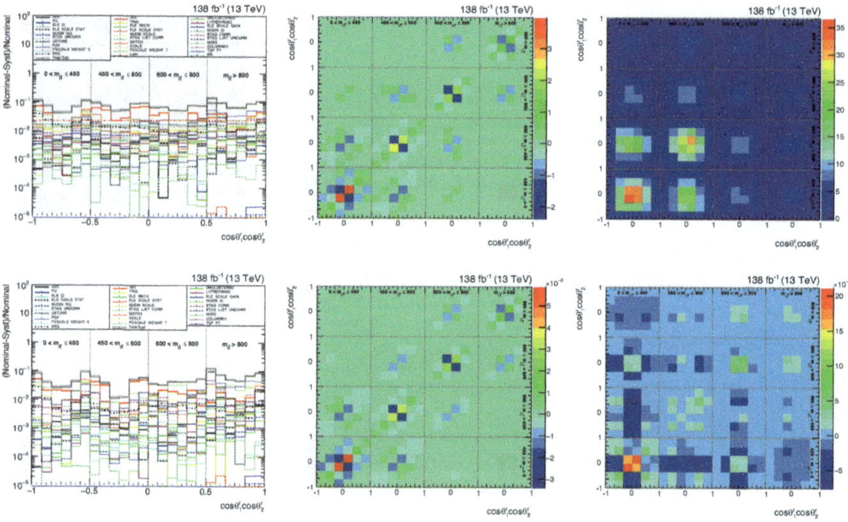

Fig. 7.48 Unfolded-cross section breakdown of uncertainties (Top Left), unfolded cross-section statistical covariance matrix (Top Middle), unfolded cross-section total systematic covariance matrix (Top Right), normalized unfolded-cross section breakdown of uncertainties (Bottom Left), normalized unfolded cross-section statistical covariance matrix (Bottom Middle), normalized unfolded cross-section total systematic covariance matrix (Bottom Right) for diagonal spin correlation observable $\cos\theta_1^r \cos\theta_2^r$ as a function of $m_{t\bar{t}}$

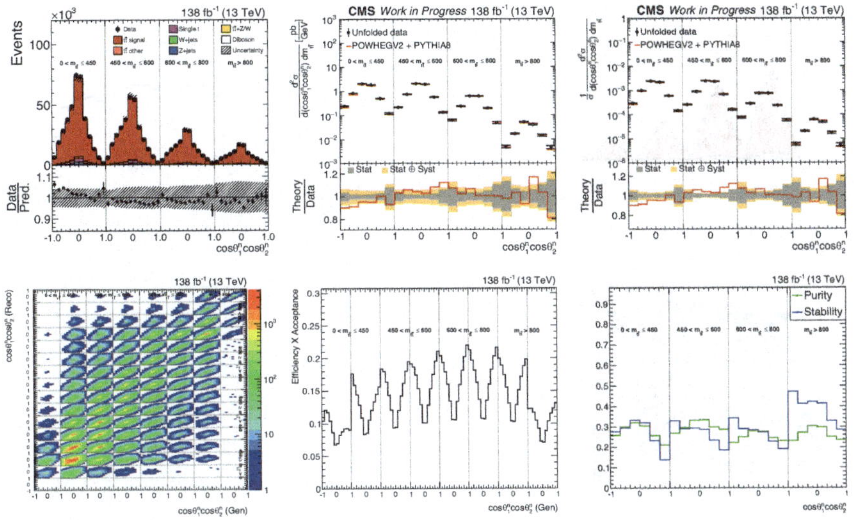

Fig. 7.49 Reconstructed detector-level distribution (Top Left), absolute cross-section unfolded to parton-level (Top Middle), normalized cross-section unfolded to parton-level (Top Right), detector response-matrix (Bottom Left), efficiency×acceptance (Bottom Middle), Purity and Stability (Bottom Right) for diagonal spin correlation observable $\cos\theta_1^n \cos\theta_2^n$ as a function of $m_{t\bar{t}}$, from which differential spin-density coefficients C_{nn} (sensitive to spin-density coefficient functions c_{nn}) are extracted

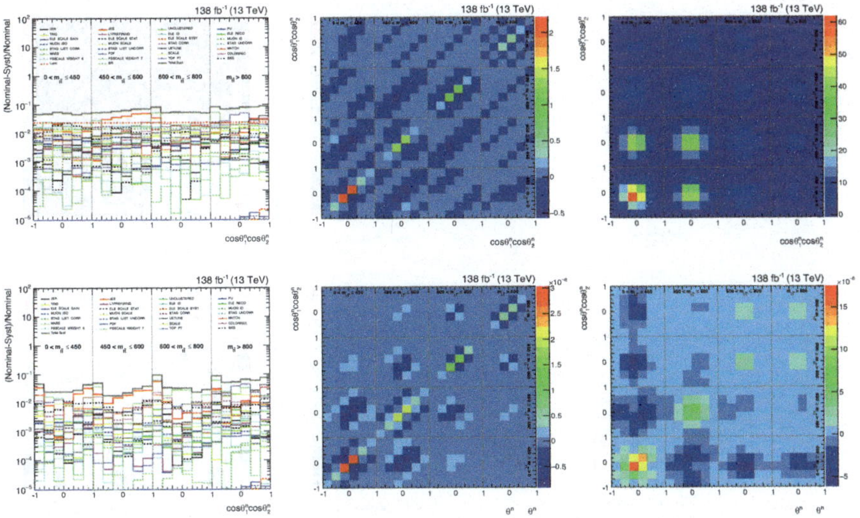

Fig. 7.50 Unfolded-cross section breakdown of uncertainties (Top Left), unfolded cross-section statistical covariance matrix (Top Middle), unfolded cross-section total systematic covariance matrix (Top Right), normalized unfolded-cross section breakdown of uncertainties (Bottom Left), normalized unfolded cross-section statistical covariance matrix (Bottom Middle), normalized unfolded cross-section total systematic covariance matrix (Bottom Right) for diagonal spin correlation observable $\cos\theta_1^n \cos\theta_2^n$ as a function of $m_{t\bar{t}}$

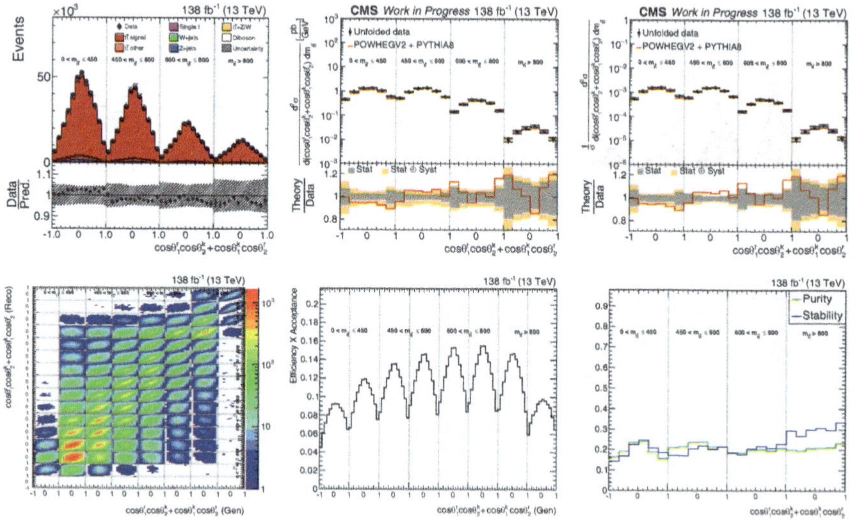

Fig. 7.51 Reconstructed detector-level distribution (Top Left), absolute cross-section unfolded to parton-level (Top Middle), normalized cross-section unfolded to parton-level (Top Right), detector response-matrix (Bottom Left), efficiency×acceptance (Bottom Middle), Purity and Stability (Bottom Right) for off-diagonal spin correlation sum observable $\cos\theta_1^r \cos\theta_2^k + \cos\theta_1^k \cos\theta_2^r$ as a function of $m_{t\bar{t}}$, from which differential spin-density coefficients $C_{rk} + C_{kr}$ (sensitive to spin-density coefficient functions c_{rk}) are extracted

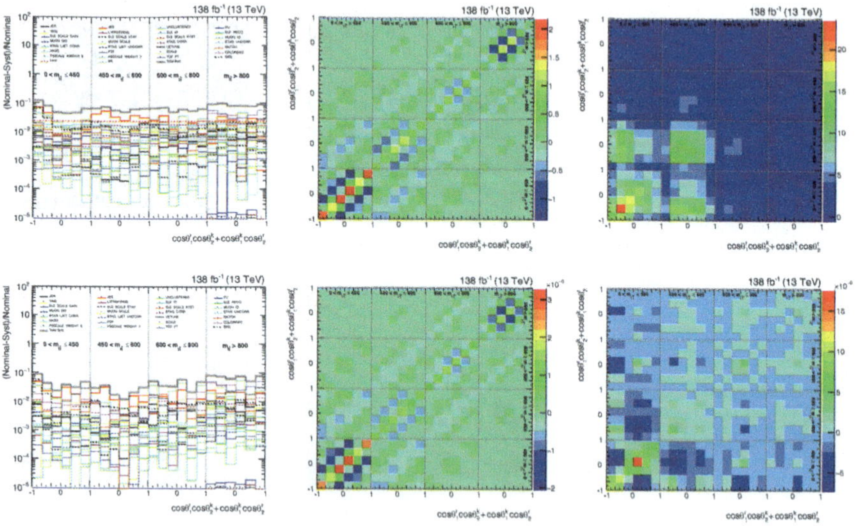

Fig. 7.52 Unfolded-cross section breakdown of uncertainties (Top Left), unfolded cross-section statistical covariance matrix (Top Middle), unfolded cross-section total systematic covariance matrix (Top Right), normalized unfolded-cross section breakdown of uncertainties (Bottom Left), normalized unfolded cross-section statistical covariance matrix (Bottom Middle), normalized unfolded cross-section total systematic covariance matrix (Bottom Right) for off-diagonal spin correlation sum observable $\cos\theta_1^r \cos\theta_2^k + \cos\theta_1^k \cos\theta_2^r$ as a function of $m_{t\bar{t}}$

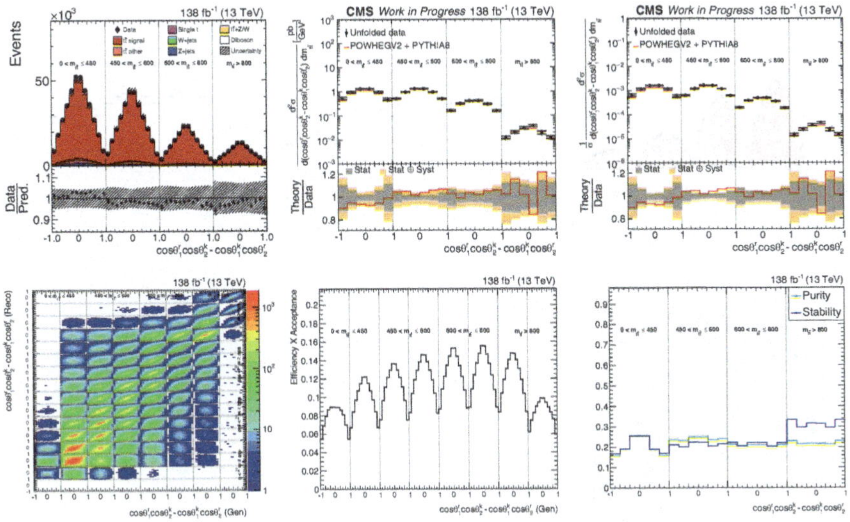

Fig. 7.53 Reconstructed detector-level distribution (Top Left), absolute cross-section unfolded to parton-level (Top Middle), normalized cross-section unfolded to parton-level (Top Right), detector response-matrix (Bottom Left), efficiency×acceptance (Bottom Middle), Purity and Stability (Bottom Right) for off-diagonal spin correlation difference observable $\cos\theta_1^r \cos\theta_2^k - \cos\theta_1^k \cos\theta_2^r$ as a function of $m_{t\bar{t}}$, from which differential spin-density coefficients $C_{rk} - C_{kr}$ (sensitive to spin-density coefficient functions c_n) are extracted

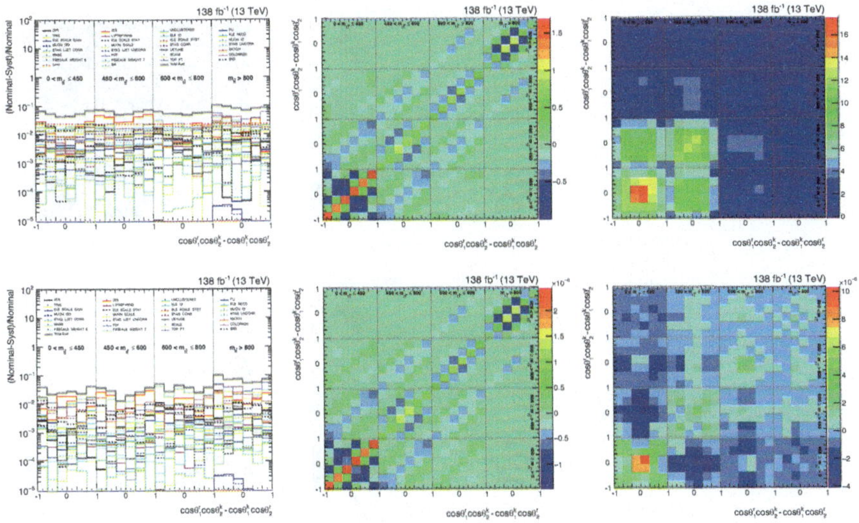

Fig. 7.54 Unfolded-cross section breakdown of uncertainties (Top Left), unfolded cross-section statistical covariance matrix (Top Middle), unfolded cross-section total systematic covariance matrix (Top Right), normalized unfolded-cross section breakdown of uncertainties (Bottom Left), normalized unfolded cross-section statistical covariance matrix (Bottom Middle), normalized unfolded cross-section total systematic covariance matrix (Bottom Right) for off-diagonal spin correlation difference observable $\cos\theta_1^r \cos\theta_2^k - \cos\theta_1^k \cos\theta_2^r$ as a function of $m_{t\bar{t}}$

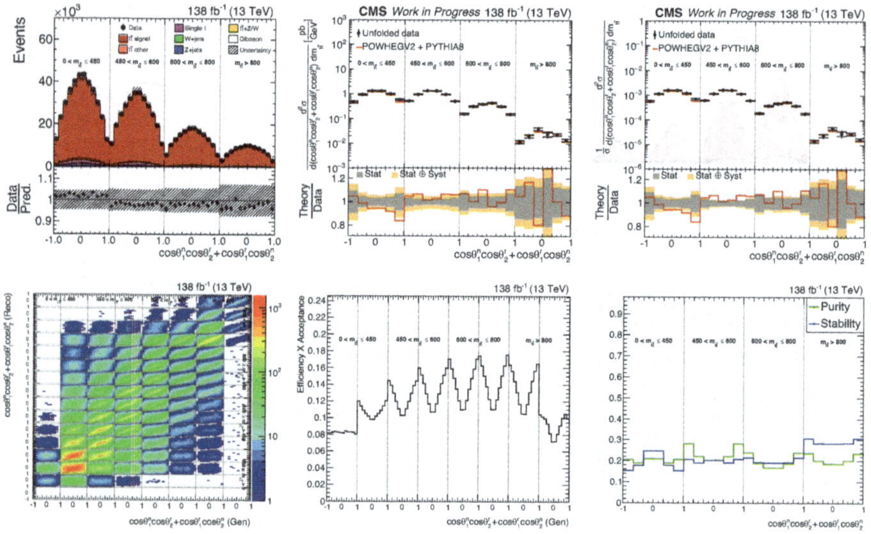

Fig. 7.55 Reconstructed detector-level distribution (Top Left), absolute cross-section unfolded to parton-level (Top Middle), normalized cross-section unfolded to parton-level (Top Right), detector response-matrix (Bottom Left), efficiency×acceptance (Bottom Middle), Purity and Stability (Bottom Right) for off-diagonal spin correlation sum observable $\cos\theta_1^n \cos\theta_2^r + \cos\theta_1^r \cos\theta_2^n$ as a function of $m_{t\bar{t}}$, from which differential spin-density coefficients $C_{nr} + C_{rn}$ (sensitive to spin-density coefficient functions c_{nr}) are extracted

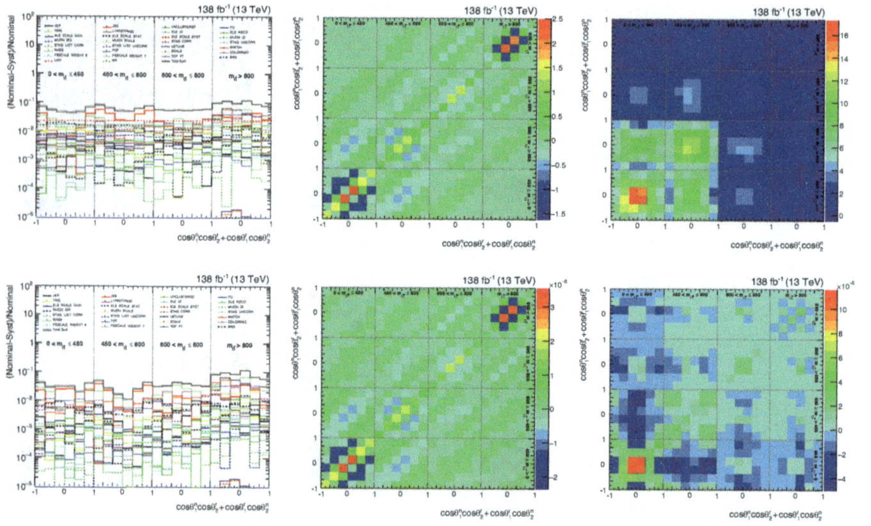

Fig. 7.56 Unfolded-cross section breakdown of uncertainties (Top Left), unfolded cross-section statistical covariance matrix (Top Middle), unfolded cross-section total systematic covariance matrix (Top Right), normalized unfolded-cross section breakdown of uncertainties (Bottom Left), normalized unfolded cross-section statistical covariance matrix (Bottom Middle), normalized unfolded cross-section total systematic covariance matrix (Bottom Right) for off-diagonal spin correlation sum observable $\cos\theta_1^n \cos\theta_2^r + \cos\theta_1^r \cos\theta_2^n$ as a function of $m_{t\bar{t}}$

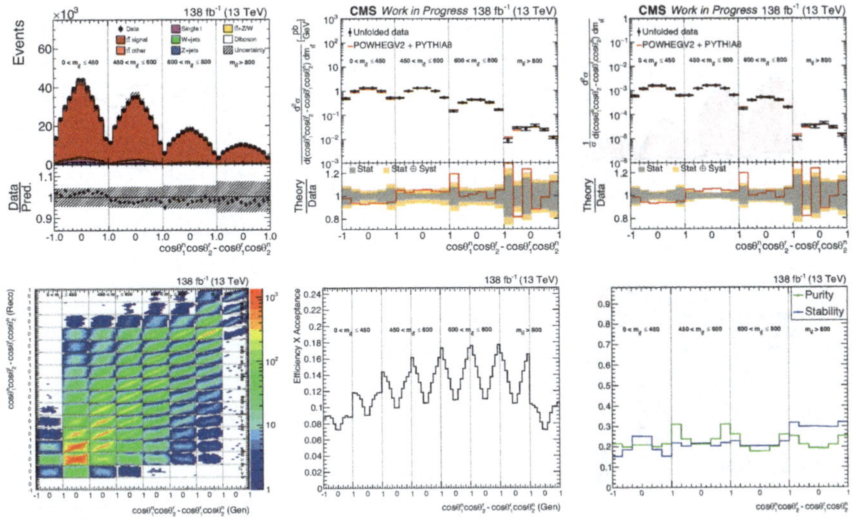

Fig. 7.57 Reconstructed detector-level distribution (Top Left), absolute cross-section unfolded to parton-level (Top Middle), normalized cross-section unfolded to parton-level (Top Right), detector response-matrix (Bottom Left), efficiency×acceptance (Bottom Middle), Purity and Stability (Bottom Right) for off-diagonal spin correlation difference observable $\cos\theta_1^n \cos\theta_2^r - \cos\theta_1^r \cos\theta_2^n$ as a function of $m_{t\bar{t}}$, from which differential spin-density coefficients $C_{nr} - C_{rn}$ (sensitive to spin-density coefficient functions c_k) are extracted

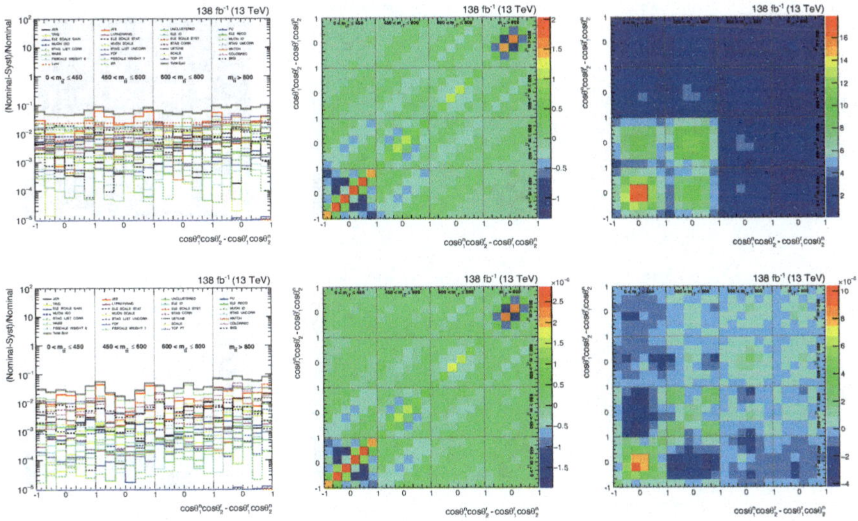

Fig. 7.58 Unfolded-cross section breakdown of uncertainties (Top Left), unfolded cross-section statistical covariance matrix (Top Middle), unfolded cross-section total systematic covariance matrix (Top Right), normalized unfolded-cross section breakdown of uncertainties (Bottom Left), normalized unfolded cross-section statistical covariance matrix (Bottom Middle), normalized unfolded cross-section total systematic covariance matrix (Bottom Right) for off-diagonal spin correlation difference observable $\cos\theta_1^n \cos\theta_2^r - \cos\theta_1^r \cos\theta_2^n$ as a function of $m_{t\bar{t}}$

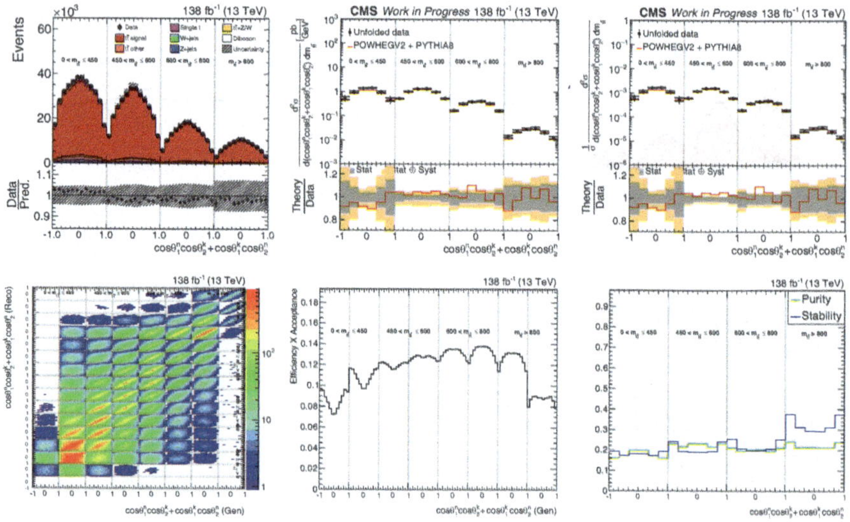

Fig. 7.59 Reconstructed detector-level distribution (Top Left), absolute cross-section unfolded to parton-level (Top Middle), normalized cross-section unfolded to parton-level (Top Right), detector response-matrix (Bottom Left), efficiency×acceptance (Bottom Middle), Purity and Stability (Bottom Right) for off-diagonal spin correlation sum observable $\cos\theta_1^n \cos\theta_2^k + \cos\theta_1^k \cos\theta_2^n$ as a function of $m_{t\bar{t}}$, from which differential spin-density coefficients $C_{nk} + C_{kn}$ (sensitive to spin-density coefficient functions c_{kn}) are extracted

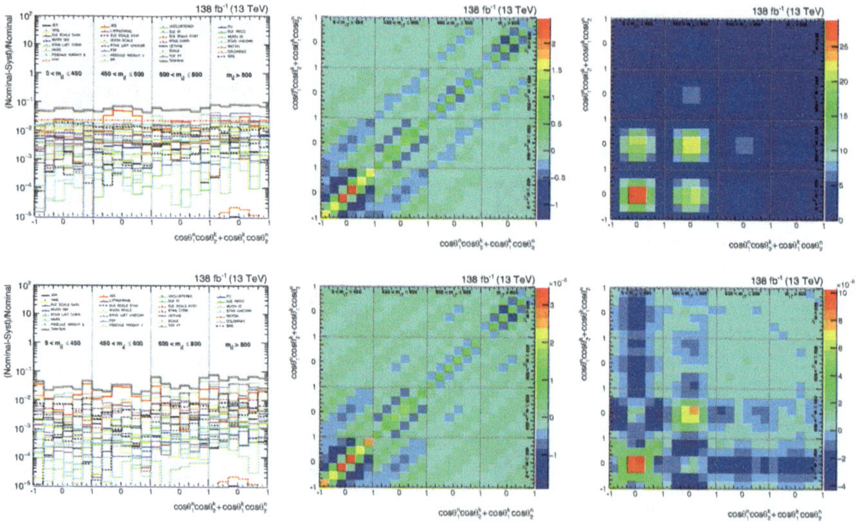

Fig. 7.60 Unfolded-cross section breakdown of uncertainties (Top Left), unfolded cross-section statistical covariance matrix (Top Middle), unfolded cross-section total systematic covariance matrix (Top Right), normalized unfolded-cross section breakdown of uncertainties (Bottom Left), normalized unfolded cross-section statistical covariance matrix (Bottom Middle), normalized unfolded cross-section total systematic covariance matrix (Bottom Right) for off-diagonal spin correlation sum observable $\cos\theta_1^n \cos\theta_2^k + \cos\theta_1^k \cos\theta_2^n$ as a function of $m_{t\bar{t}}$

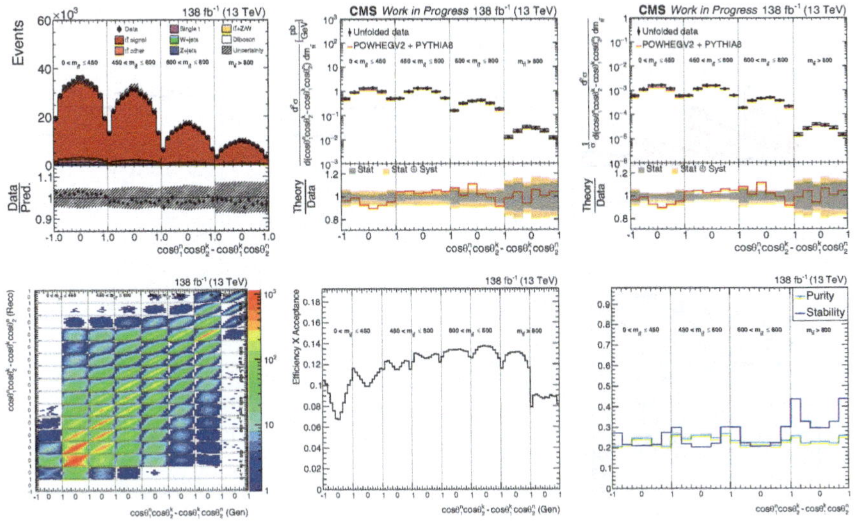

Fig. 7.61 Reconstructed detector-level distribution (Top Left), absolute cross-section unfolded to parton-level (Top Middle), normalized cross-section unfolded to parton-level (Top Right), detector response-matrix (Bottom Left), efficiency×acceptance (Bottom Middle), Purity and Stability (Bottom Right) for off-diagonal spin correlation difference observable $\cos\theta_1^n \cos\theta_2^k - \cos\theta_1^k \cos\theta_2^n$ as a function of $m_{t\bar{t}}$, from which differential spin-density coefficients $C_{nk} - C_{kn}$ (sensitive to spin-density coefficient functions $-c_r$) are extracted

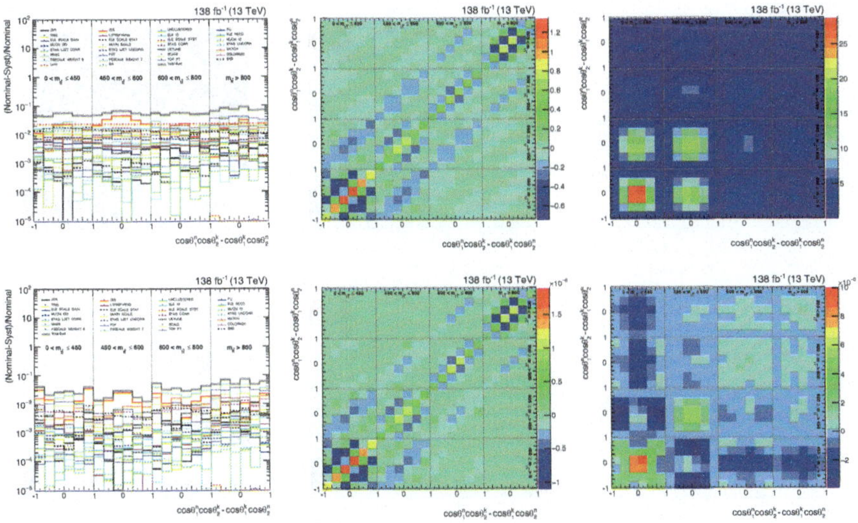

Fig. 7.62 Unfolded-cross section breakdown of uncertainties (Top Left), unfolded cross-section statistical covariance matrix (Top Middle), unfolded cross-section total systematic covariance matrix (Top Right), normalized unfolded-cross section breakdown of uncertainties (Bottom Left), normalized unfolded cross-section statistical covariance matrix (Bottom Middle), normalized unfolded cross-section total systematic covariance matrix (Bottom Right) for off-diagonal spin correlation difference observable $\cos\theta_1^n \cos\theta_2^k - \cos\theta_1^k \cos\theta_2^n$ as a function of $m_{t\bar{t}}$

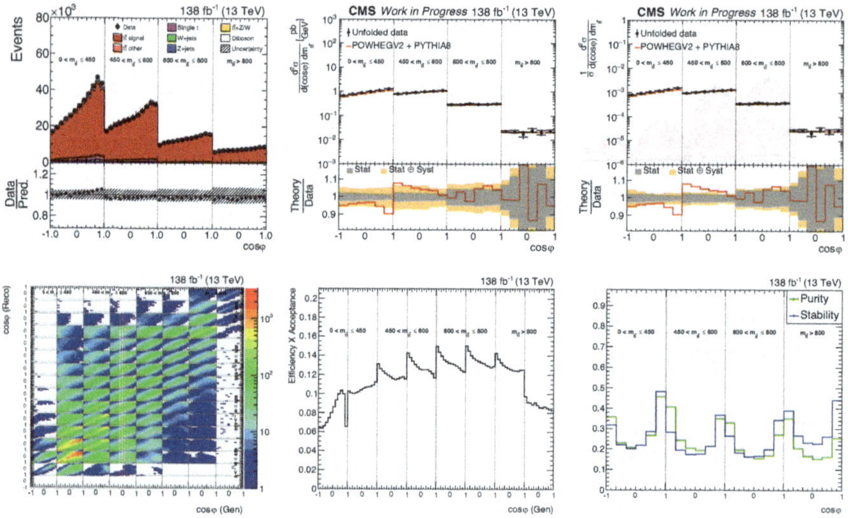

Fig. 7.63 Reconstructed detector-level distribution (Top Left), absolute cross-section unfolded to parton-level (Top Middle), normalized cross-section unfolded to parton-level (Top Right), detector response-matrix (Bottom Left), efficiency×acceptance (Bottom Middle), Purity and Stability (Bottom Right) for observable $\cos\varphi$ as a function of $m_{t\bar{t}}$, from which differential spin-density coefficients $D = -\frac{1}{3}(C_{kk} + C_{rr} + C_{nn})$ (sensitive to spin-density coefficient functions $-\frac{1}{3}(c_{kk} + c_{rr} + c_{nn})$) are extracted

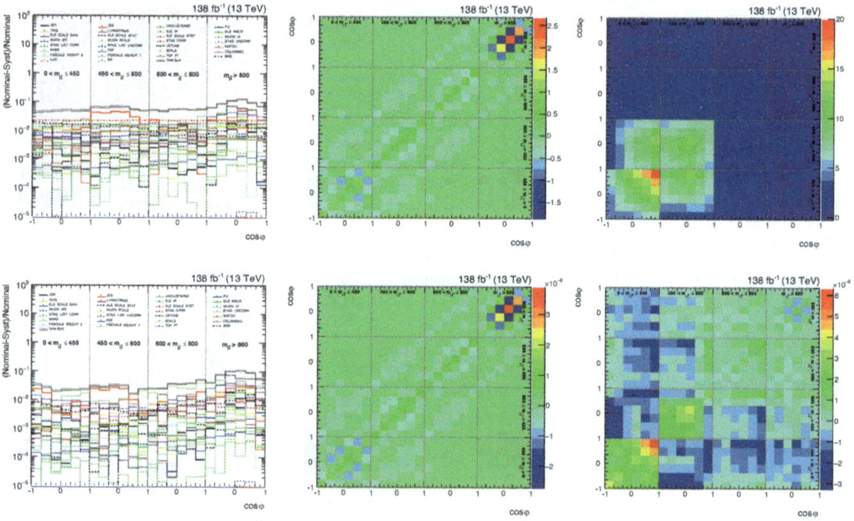

Fig. 7.64 Unfolded-cross section breakdown of uncertainties (Top Left), unfolded cross-section statistical covariance matrix (Top Middle), unfolded cross-section total systematic covariance matrix (Top Right), normalized unfolded-cross section breakdown of uncertainties (Bottom Left), normalized unfolded cross-section statistical covariance matrix (Bottom Middle), normalized unfolded cross-section total systematic covariance matrix (Bottom Right) for observable $\cos\varphi$ as a function of $m_{t\bar{t}}$

7.2 Extracted Top Quark Polarization and $t\bar{t}$ Spin Correlation Coefficients

The set[2] of one-dimensional spin density coefficients extracted from the normalized cross-sections unfolded to parton-level (as described in Sect. 6.6) is shown in Table 7.1, and the two-dimensional set, as a function of $m_{t\bar{t}}$, is shown in Tables 7.2, 7.3, and 7.4.

The extracted coefficients are compared with predictions from the POWHEGV2 +PYTHIA8 MC simulation and the one-dimensional set are additionally compared to calculations for $t\bar{t}$ production at NLO in QCD with electroweak corrections [2]. There is generally decent agreement between the measured coefficients and the predictions, with better agreement when compared to perturbative QCD calculations at NLO. Due to significantly increased luminosity and several optimizations that reduced background contributions and systematic uncertainties, the measurement precision for extracted coefficients is improved compared to CMS TOP-18-006 [1], by as much as a factor of two for some coefficients. The only remarkable observed deviations from predictions were for the differential measurements of B_1^k, which may be explained by b-quark fragmentation having a large effect on top quark momentum in the $t\bar{t}$ ZMF.

[2] A summary of all spin density observables, their respective coefficients, and which coefficient functions of the spin density matrix they probe is provided in Table 6.1.

Table 7.1 Top quark polarization and $t\bar{t}$ spin correlation coefficients extracted from the normalized differential cross-sections with combined statistical and systematic uncertainties. Predicted values from the POWHEGV2+PYTHIA8 MC simulation are quoted with combined statistical and scale uncertainties. NLO calculated values are quoted with scale uncertainties [2]

Coefficient	Measured	POWHEGV2	NLO calculation
B_1^k	-0.029 ± 0.019	$0.005^{+0.000}_{-0.000}$	$4.0^{+1.7}_{-1.2} \times 10^{-3}$
B_2^k	-0.017 ± 0.019	$0.005^{+0.000}_{-0.000}$	$4.0^{+1.7}_{-1.2} \times 10^{-3}$
B_1^r	-0.004 ± 0.011	$0.002^{+0.000}_{-0.000}$	$1.6^{+1.2}_{-0.9} \times 10^{-3}$
B_2^r	-0.009 ± 0.011	$0.002^{+0.000}_{-0.000}$	$1.6^{+1.2}_{-0.9} \times 10^{-3}$
B_1^n	0.002 ± 0.004	$-0.000^{+0.000}_{-0.000}$	$5.7^{+0.5}_{-0.4} \times 10^{-3}$
B_2^n	0.008 ± 0.004	$-0.000^{+0.000}_{-0.000}$	$5.7^{+0.5}_{-0.4} \times 10^{-3}$
C_{kk}	0.351 ± 0.020	$0.308^{+0.005}_{-0.005}$	$0.321^{+0.002}_{-0.002}$
C_{rr}	0.110 ± 0.019	$0.042^{+0.008}_{-0.006}$	$0.071^{+0.008}_{-0.006}$
C_{nn}	0.345 ± 0.018	$0.317^{+0.001}_{-0.001}$	$0.326^{+0.002}_{-0.002}$
$C_{rk} + C_{kr}$	-0.233 ± 0.060	$-0.206^{+0.004}_{-0.004}$	$-0.206^{+0.002}_{-0.002}$
$C_{rk} - C_{kr}$	-0.007 ± 0.016	$0.001^{+0.000}_{-0.000}$	0
$C_{nr} + C_{rn}$	-0.015 ± 0.015	$0.000^{+0.000}_{-0.000}$	$1.06^{+0.01}_{-0.01} \times 10^{-3}$
$C_{nr} - C_{rn}$	-0.018 ± 0.012	$0.000^{+0.000}_{-0.000}$	0
$C_{nk} + C_{kn}$	0.010 ± 0.015	$0.000^{+0.000}_{-0.000}$	$2.15^{+0.04}_{-0.07} \times 10^{-3}$
$C_{nk} - C_{kn}$	0.008 ± 0.011	$-0.001^{+0.000}_{-0.000}$	0
D	-0.254 ± 0.009	$-0.222^{+0.003}_{-0.004}$	$-0.243^{+0.003}_{-0.003}$

Table 7.2 Differential top quark polarization coefficients, as a function of $m_{t\bar{t}}$, extracted from the normalized differential cross-sections with combined statistical and systematic uncertainties. Predicted values from the POWHEGV2+PYTHIA8 MC simulation are quoted with combined statistical and scale uncertainties

Coefficient	Measured	POWHEGV2
B_1^k, $250 < m_{t\bar{t}} < 450$	-0.046 ± 0.016	$0.006^{+0.000}_{-0.000}$
B_1^k, $450 < m_{t\bar{t}} < 600$	0.027 ± 0.030	$0.005^{+0.000}_{-0.000}$
B_1^k, $600 < m_{t\bar{t}} < 800$	-0.101 ± 0.033	$0.004^{+0.000}_{-0.000}$
B_1^k, $800 < m_{t\bar{t}} < 2000$	-0.001 ± 0.049	$0.002^{+0.000}_{-0.000}$
B_2^k, $250 < m_{t\bar{t}} < 450$	-0.009 ± 0.015	$0.006^{+0.000}_{-0.000}$
B_2^k, $450 < m_{t\bar{t}} < 600$	-0.001 ± 0.030	$0.005^{+0.000}_{-0.000}$
B_2^k, $600 < m_{t\bar{t}} < 800$	-0.046 ± 0.034	$0.004^{+0.000}_{-0.000}$
B_2^k, $800 < m_{t\bar{t}} < 2000$	-0.002 ± 0.052	$0.002^{+0.000}_{-0.000}$
B_1^r, $250 < m_{t\bar{t}} < 450$	-0.001 ± 0.029	$0.003^{+0.000}_{-0.000}$
B_1^r, $450 < m_{t\bar{t}} < 600$	0.004 ± 0.018	$0.002^{+0.000}_{-0.000}$
B_1^r, $600 < m_{t\bar{t}} < 800$	-0.013 ± 0.023	$0.001^{+0.000}_{-0.000}$
B_1^r, $800 < m_{t\bar{t}} < 2000$	-0.041 ± 0.029	$0.001^{+0.000}_{-0.000}$
B_2^r, $250 < m_{t\bar{t}} < 450$	-0.021 ± 0.028	$0.003^{+0.000}_{-0.000}$
B_2^r, $450 < m_{t\bar{t}} < 600$	-0.007 ± 0.020	$0.002^{+0.000}_{-0.000}$
B_2^r, $600 < m_{t\bar{t}} < 800$	0.006 ± 0.022	$0.001^{+0.000}_{-0.000}$
B_2^r, $800 < m_{t\bar{t}} < 2000$	-0.026 ± 0.030	$-\ 0.000^{+0.000}_{-0.000}$
B_1^n, $250 < m_{t\bar{t}} < 450$	0.018 ± 0.019	$-\ 0.000^{+0.000}_{-0.000}$
B_1^n, $450 < m_{t\bar{t}} < 600$	-0.007 ± 0.015	$-\ 0.000^{+0.000}_{-0.000}$
B_1^n, $600 < m_{t\bar{t}} < 800$	0.002 ± 0.020	$-\ 0.000^{+0.000}_{-0.000}$
B_1^n, $800 < m_{t\bar{t}} < 2000$	-0.003 ± 0.021	$-\ 0.000^{+0.000}_{-0.000}$
B_2^n, $250 < m_{t\bar{t}} < 450$	-0.001 ± 0.017	$0.000^{+0.000}_{-0.000}$
B_2^n, $450 < m_{t\bar{t}} < 600$	0.025 ± 0.014	$-\ 0.000^{+0.000}_{-0.000}$
B_2^n, $600 < m_{t\bar{t}} < 800$	-0.010 ± 0.019	$0.000^{+0.000}_{-0.000}$
B_2^n, $800 < m_{t\bar{t}} < 2000$	0.008 ± 0.022	$-\ 0.000^{+0.000}_{-0.000}$

Table 7.3 Differential $t\bar{t}$ diagonal spin correlation coefficients, as a function of $m_{t\bar{t}}$, extracted from the normalized differential cross-sections with combined statistical and systematic uncertainties. Predicted values from the POWHEGV2+PYTHIA8 MC simulation are quoted with combined statistical and scale uncertainties

Coefficient	Measured	POWHEGV2
C_{kk}, $250 < m_{t\bar{t}} < 450$	0.475 ± 0.074	$0.450^{+0.004}_{-0.005}$
C_{kk}, $450 < m_{t\bar{t}} < 600$	0.299 ± 0.042	$0.279^{+0.004}_{-0.004}$
C_{kk}, $600 < m_{t\bar{t}} < 800$	0.126 ± 0.040	$0.120^{+0.010}_{-0.008}$
C_{kk}, $800 < m_{t\bar{t}} < 2000$	-0.031 ± 0.056	$-0.041^{+0.001}_{-0.001}$
C_{rr}, $250 < m_{t\bar{t}} < 450$	0.271 ± 0.043	$0.188^{+0.008}_{-0.007}$
C_{rr}, $450 < m_{t\bar{t}} < 600$	-0.020 ± 0.039	$-0.035^{+0.008}_{-0.006}$
C_{rr}, $600 < m_{t\bar{t}} < 800$	-0.018 ± 0.056	$-0.115^{+0.010}_{-0.008}$
C_{rr}, $800 < m_{t\bar{t}} < 2000$	-0.102 ± 0.069	$-0.133^{+0.001}_{-0.001}$
C_{nn}, $250 < m_{t\bar{t}} < 450$	0.487 ± 0.027	$0.420^{+0.002}_{-0.002}$
C_{nn}, $450 < m_{t\bar{t}} < 600$	0.264 ± 0.038	$0.265^{+0.001}_{-0.001}$
C_{nn}, $600 < m_{t\bar{t}} < 800$	0.223 ± 0.043	$0.206^{+0.003}_{-0.004}$
C_{nn}, $800 < m_{t\bar{t}} < 2000$	0.139 ± 0.044	$0.169^{+0.001}_{-0.001}$
D, $250 < m_{t\bar{t}} < 450$	-0.383 ± 0.019	$-0.353^{+0.005}_{-0.005}$
D, $450 < m_{t\bar{t}} < 600$	-0.180 ± 0.026	$-0.170^{+0.003}_{-0.004}$
D, $600 < m_{t\bar{t}} < 800$	-0.101 ± 0.029	$-0.070^{+0.004}_{-0.005}$
D, $800 < m_{t\bar{t}} < 2000$	-0.005 ± 0.026	$0.002^{+0.000}_{-0.000}$

Table 7.4 Differential $t\bar{t}$ cross spin correlation coefficients, as a function of $m_{t\bar{t}}$, extracted from the normalized differential cross-sections with combined statistical and systematic uncertainties. Predicted values from the POWHEGV2+PYTHIA8 MC simulation are quoted with combined statistical and scale uncertainties

Coefficient	Measured	POWHEGV2
$C_{rk} + C_{kr}$, $250 < m_{t\bar{t}} < 450$	-0.301 ± 0.129	$-0.202^{+0.004}_{-0.004}$
$C_{rk} + C_{kr}$, $450 < m_{t\bar{t}} < 600$	-0.182 ± 0.054	$-0.224^{+0.004}_{-0.003}$
$C_{rk} + C_{kr}$, $600 < m_{t\bar{t}} < 800$	-0.249 ± 0.076	$-0.199^{+0.006}_{-0.004}$
$C_{rk} + C_{kr}$, $800 < m_{t\bar{t}} < 2000$	-0.130 ± 0.087	$-0.149^{+0.001}_{-0.006}$
$C_{rk} - C_{kr}$, $250 < m_{t\bar{t}} < 450$	-0.063 ± 0.058	$0.001^{+0.000}_{-0.000}$
$C_{rk} - C_{kr}$, $450 < m_{t\bar{t}} < 600$	0.039 ± 0.048	$0.001^{+0.001}_{-0.001}$
$C_{rk} - C_{kr}$, $600 < m_{t\bar{t}} < 800$	0.003 ± 0.066	$0.001^{+0.001}_{-0.001}$
$C_{rk} - C_{kr}$, $800 < m_{t\bar{t}} < 2000$	-0.086 ± 0.077	$0.002^{+0.001}_{-0.001}$
$C_{nr} + C_{rn}$, $250 < m_{t\bar{t}} < 450$	-0.051 ± 0.053	$0.001^{+0.000}_{-0.000}$
$C_{nr} + C_{rn}$, $450 < m_{t\bar{t}} < 600$	-0.010 ± 0.043	$0.000^{+0.001}_{-0.001}$
$C_{nr} + C_{rn}$, $600 < m_{t\bar{t}} < 800$	-0.003 ± 0.056	$0.000^{+0.001}_{-0.001}$
$C_{nr} + C_{rn}$, $800 < m_{t\bar{t}} < 2000$	0.022 ± 0.072	$0.001^{+0.001}_{-0.001}$
$C_{nr} - C_{rn}$, $250 < m_{t\bar{t}} < 450$	0.006 ± 0.069	$0.000^{+0.000}_{-0.000}$
$C_{nr} - C_{rn}$, $450 < m_{t\bar{t}} < 600$	-0.015 ± 0.041	$0.001^{+0.001}_{-0.001}$
$C_{nr} - C_{rn}$, $600 < m_{t\bar{t}} < 800$	-0.049 ± 0.060	$0.001^{+0.001}_{-0.001}$
$C_{nr} - C_{rn}$, $800 < m_{t\bar{t}} < 2000$	-0.014 ± 0.070	$-0.001^{+0.001}_{-0.001}$
$C_{nk} + C_{kn}$, $250 < m_{t\bar{t}} < 450$	0.082 ± 0.066	$0.000^{+0.000}_{-0.001}$
$C_{nk} + C_{kn}$, $450 < m_{t\bar{t}} < 600$	0.013 ± 0.039	$-0.000^{+0.001}_{-0.001}$
$C_{nk} + C_{kn}$, $600 < m_{t\bar{t}} < 800$	-0.062 ± 0.054	$0.001^{+0.001}_{-0.001}$
$C_{nk} + C_{kn}$, $800 < m_{t\bar{t}} < 2000$	0.026 ± 0.060	$-0.000^{+0.001}_{-0.001}$
$C_{nk} - C_{kn}$, $250 < m_{t\bar{t}} < 450$	0.018 ± 0.039	$-0.000^{+0.000}_{-0.000}$
$C_{nk} - C_{kn}$, $450 < m_{t\bar{t}} < 600$	0.017 ± 0.032	$-0.001^{+0.001}_{-0.001}$
$C_{nk} - C_{kn}$, $600 < m_{t\bar{t}} < 800$	-0.066 ± 0.048	$-0.000^{+0.001}_{-0.001}$
$C_{nk} - C_{kn}$, $800 < m_{t\bar{t}} < 2000$	0.112 ± 0.056	$-0.001^{+0.001}_{-0.001}$

References

1. CMS Collaboration, Measurement of the top quark polarization and $t\bar{t}$ spin correlations using dilepton final states in proton-proton collisions at $\sqrt{s} = 13\,\text{TeV}$. Phys. Rev. D **100**(7), 072002 (2019). Preliminary results. https://doi.org/10.1103/PhysRevD.100.072002. arXiv: 1907.03729. [Online]. Available: https://cds.cern.ch/record/2681777
2. W. Bernreuther, D. Heisler, Z.-G. Si, A set of top quark spin correlation and polarization observables for the LHC: standard model predictions and new physics contributions. J. High Energy Phys. **2015** (2015). https://doi.org/10.1007/JHEP12(2015)026

Chapter 8
Summary and Outlook

Abstract This final chapter provides a summary of the key findings from the measurement results presented in this dissertation and their implications for the field of particle physics. It reflects on the progress made in measuring $t\bar{t}$ spin correlations as a precision test of the SM. The potential for future research by expanding scope of these measurements is considered. The dissertation concludes with an outlook on the future of $t\bar{t}$ spin correlation measurements at the anticipated HL-LHC with upgraded detector technology, as well as at proposed lepton colliders.

This dissertation presents the first differential measurements of all the independent coefficients of the top-spin components of the $t\bar{t}$ production density matrix. The measurements were conducted using $137.7\text{fb}^{-1} \pm 1.6\%$ of data collected by the CMS experiment at the LHC with $\sqrt{s} = 13\,\text{TeV}$ during 2016, 2017, and 2018. All channels (e^+e^-, $e^\pm\mu^\mp$, $\mu^+\mu^-$) of the $t\bar{t}$ dileptonic decay mode with final states containing two oppositely charged leptons were targeted; 16 observables were probed with a total of 80 differential measurements. The measured observables were corrected for detector efficiencies, acceptances, and migrations, unfolded to parton-level, and extrapolated to the full phase space using a regularized unfolding procedure with detector response obtained from MC simulated SM predictions with NLO matrix element accuracy interfaced with parton-shower and hadronization algorithms. The spin density coefficients were extracted from the unfolded distributions and compared to MC simulations and theoretical predictions, constituting a precision test of the SM. With significantly increased luminosity, several optimizations that reduce background contributions and systematic uncertainties, and the inclusion of $t\bar{t}$ events the via τ decays as signal, signal purity has been increased from $\sim$ 79 to $\sim$ 93% and measurement precision for one-dimensional normalized unfolded cross-sections and extracted coefficients were improved by as much as a factor of two compared to CMS TOP-18-006 [1]. The measured coefficients showed decent agreement with the MC predictions, but the agreement was improved when compared to QCD perturbative calculations for $t\bar{t}$ production

J. R. Thieman, *Top Quark Spin Correlations with the CMS Detector*,
Springer Theses, https://doi.org/10.1007/978-3-031-76117-1_8

at NLO with electroweak corrections. The only notable deviations from predictions were observed for the differential measurements of coefficient B_1^k, which may be partially explained by b-quark fragmentation having a significant effect on top quark momentum in the $t\bar{t}$ ZMF.

This dissertation presents a set of measurements that can be further used for various interpretations. For example, a SMEFT interpretation can use these measurements to set limits on dimension-six operators relevant for hadronic $t\bar{t}$ production [1]. Additionally, the measurements can be used for a SUSY interpretation to set limits on light supersymmetric top squark production or to probe for entanglement [2] or a violation of Bell's Inequality in the $t\bar{t}$ system [3]. Furthermore, the measurements can be used to search for evidence of toponium near $t\bar{t}$ production threshold [4]. To enable a SMEFT interpretation and fitting of multiple distributions, the top quark research group at Purdue University is determining the statistical and systematic correlation matrices for all bins of all differential cross sections. The statistical correlations are estimated using bootstrap resampling of the data, while the systematic correlations are estimated by simultaneously evaluating the systematic variations for all measured bins.

There are various ways to expand measurements of the top-spin components of the $t\bar{t}$ production density matrix. In addition to two-dimensional differential measurements based on the $t\bar{t}$ invariant mass, differential measurements based on the scattering angle of the top quark in the $t\bar{t}$ ZMF and the p_{T} of the top quark are also highly motivated. Ideally, three-dimensional differential measurements that consider both the $t\bar{t}$ invariant mass and top quark scattering angle in the $t\bar{t}$ ZMF would offer the most effective way to probe perturbative QCD predictions for $t\bar{t}$ production. Furthermore, lab frame observables such as $|\Delta\phi_{\ell\bar{\ell}}|$ and $|\Delta\eta_{\ell\bar{\ell}}|$ (the difference in azimuthal angle ϕ and the difference in pseudorapidity η between the two leptons) are indirectly sensitive to the spin density coefficients and can be measured with excellent experimental resolution.

In the future, the $t\bar{t}$ production density matrix spin coefficients can be measured with the CMS 13.6 TeV data set, and significantly improved precision at the HL-LHC. The HL-LHC will provide substantially higher luminosity and upgraded detectors that will improve measurement capabilities and reduce systematic uncertainties, including better object identification and resolution, allowing for more precise measurements.

The limited collision energies at lepton colliders have prohibited $t\bar{t}$ production and measurements of top quark polarizations and $t\bar{t}$ spin correlations. Conducting measurements of the $t\bar{t}$ production density matrix at the proposed International Linear Collider (ILC) or Compact Linear Collider (CLIC) would probe electroweak $t\bar{t}$ production, which, unlike QCD production, predicts significant top quark and anti-quark polarizations. Thus, measurements at lepton colliders would be complementary to those performed at hadron colliders.

References

1. CMS Collaboration, Measurement of the top quark polarization and $t\bar{t}$ spin correlations using dilepton final states in proton-proton collisions at $\sqrt{s} = 13$ TeV. Phys. Rev. D **100**(7), 072002 (2019). Preliminary results. https://doi.org/10.1103/PhysRevD.100.072002. arXiv: 1907.03729 [Online]. Available: https://cds.cern.ch/record/2681777
2. Y. Afik, J.R.M. de Nova, Entanglement and quantum tomography with top quarks at the LHC. Eur. Phys. J. Plus **136**(9) (2021). https://doi.org/10.1140/epjp/s13360-021-01902-1 [Online]. Available: https://doi.org/10.1140%2Fepjp%2Fs13360-021-01902-1
3. J.A. Aguilar-Saavedra, J.A. Casas, Improved tests of entanglement and bell inequalities with LHC tops. Eur. Phys. J. C **82**(8) (2022). https://doi.org/10.1140/epjc/s10052-022-10630-4 [Online]. Available: https://doi.org/10.1140%2Fepjc%2Fs10052-022-10630-4
4. B. Fuks, K. Hagiwara, K. Ma, Y.-J. Zheng, Signatures of toponium formation in lhc run 2 data. Phys. Rev. D **104**, 034023 (2021). https://doi.org/10.1103/PhysRevD.104.034023 [Online]. Available: https://link.aps.org/doi/10.1103/PhysRevD.104.034023

Appendix A
Measurements of Trigger Efficiency Scale Factors for Selection of $t\bar{t}$ Dilepton Final States

Appendix A details the method and results of measuring trigger efficiency scale factors for the selection of top quark pair ($t\bar{t}$) events that decay via the dileptonic mode with dilepton (e^+e^-, $e^\pm\mu^\mp$, $\mu^+\mu^-$) final states. The results, which were used by virtually all CMS Run II measurements analyzing $t\bar{t}$ events that decay via the dileptonic mode, are crucial for correcting the trigger efficiencies in Monte Carlo simulations to match those observed in real data. It discusses the datasets and triggers used, the criteria for object and event selection, and also covers statistical and systematic uncertainties associated with these measurements. Additionally, this appendix includes dedicated trigger efficiency scale factor measurements that were used specifically for investigating Lorentz Invariance Violation in $t\bar{t}$ production.

A.1 Overview and Motivation

Here is presented the measurements of the trigger efficiencies and corrective scale factors that were used by virtually all CMS Run II measurements analyzing $t\bar{t}$ events that decay via the dileptonic mode with e^+e^-, $e^\pm\mu^\mp$, $\mu^+\mu^-$ final states. The methods used for the measurements and the determination of systematic uncertainties were approved by the EGamma and Muon CMS Physics Object Groups because of the inclusion of electron and muon objects respectively, as well as by the CMS Top Quark Physics Analysis Group.

Trigger efficiencies measure the ability of the CMS trigger system to select potentially interesting events for further analysis. Differences between the observed trigger efficiencies of data recorded by the detector and the expected trigger efficiencies of simulated data can bias measured physics results, and these differences are corrected via trigger efficiency scale factors that are applied to the trigger efficiencies of simulated data so that they match the observed trigger efficiencies of recorded data. Trigger efficiency corrective scale factors (SF_{Trig}) are calculated as:

J. R. Thieman, *Top Quark Spin Correlations with the CMS Detector*,
Springer Theses, https://doi.org/10.1007/978-3-031-76117-1

$$\mathrm{SF}_{\mathrm{Trig}} = \frac{\mathcal{E}^{\mathrm{Data}}}{\mathcal{E}^{\mathrm{MC}}} \tag{A.1}$$

where $\mathcal{E}^{\mathrm{Data}}$ and $\mathcal{E}^{\mathrm{MC}}$ are trigger efficiencies for data and Monte Carlo simulated data, respectively.

The method used to measure the dilepton trigger efficiencies utilizes a set of cross-triggers, that are weakly correlated with the dilepton triggers, to count the number of events passing the cross-trigger and $t\bar{t}$ dilepton final state event selection criteria (denominator), and from that set count also the subset that passes the dilepton triggers (numerator). Missing transverse energy ($E_{\mathrm{T}}^{\mathrm{miss}}$) triggers are shown to be minimally correlated with the dilepton triggers, and they provide sufficient statistics for low statistical uncertainties, so they are used as the cross-triggers for these measurements. The trigger efficiencies, for both data and Monte Carlo simulated data, are thus calculated as:

$$\mathcal{E} = \frac{N_{\{\text{Selection}\} \cap \{E_{\mathrm{T}}^{\mathrm{miss}} \text{ triggers}\} \cap \{\text{Dilepton triggers}\}}}{N_{\{\text{Selection}\} \cap \{E_{\mathrm{T}}^{\mathrm{miss}} \text{ triggers}\}}} \tag{A.2}$$

where, in addition to passing the trigger requirements, events must also pass the event selection criteria for the appropriate dilepton channel.

The dilepton trigger efficiency corrective scale factors are measured, separately in the e^+e^-, $e^\pm\mu^\mp$, $\mu^+\mu^-$ channels, as one-dimensional and two-dimensional functions of leading and sub-leading lepton p_{T}. The two-dimensional scale factors are the set recommended to be used by virtually all CMS Run II measurements analyzing $t\bar{t}$ events that decay via the dileptonic mode with e^+e^-, $e^\pm\mu^\mp$, $\mu^+\mu^-$ final states.

A.2 Data Sets and Triggers

The measurements, which were originally performed using pre-ultra legacy CMS data sets but were repeated using the ultra legacy sets, are comprehensively documented, are designed to be useful for a wide range of studies, are preserved for long-term access, and include high-level physics objects such as reconstructed leptons, jets, and missing transverse energy. The data streams were collected by the CMS experiment in 2016, 2017, and 2018, and correspond to a combined integrated luminosity of $137.7\mathrm{fb}^{-1}$. The 2016 data set is separated into preVFP and postVFP, partitioning the year into the periods before and after HIPM was implemented. The lists of triggers for each of the 3 years were provided in collaboration with trigger experts of the CMS Top Quark Physics Analysis Group. Both dilepton and single-lepton triggers were used in order to maximize the trigger efficiency.

A.3 Object and Event Selection

The dilepton final state of $t\bar{t}$ events is characterized by the presence of at least two high-p_{T} isolated leptons with opposite electric charge, large missing transverse energy ($E_{\mathrm{T}}^{\mathrm{miss}}$), and two jets resulting from the hadronization of bottom quarks. The identification and reconstruction of the different objects is performed using the PF object-reconstruction algorithm, which combines information from all the raw sub-detector signals to accurately identify and reconstruct individual particle objects for each event.

A.3.1 Electrons

Electron candidates are selected from the reconstructed GSF electrons (defined in Sect. 5.2.4) with $p_{\mathrm{T}} > 15\,\mathrm{GeV}$ and $|\eta_{\mathrm{SC}}| < 2.4$, removing the barrel-endcap gap $1.4442 < |\eta_{SC}| < 1.566$. Selected electrons must pass the tight identification tracking requirements, including relative isolation, as recommended by the CMS EGamma Physics Object Group. Corrections also are applied to scale the raw energy measurements and smear the electron resolutions to match the accuracy and precision of reconstructed electrons in MC simulated data to those in recorded data.

A.3.2 Muons

PF muons selected for this measurement are required to be reconstructed as global muons (defined in Sect. 5.2.3), relatively isolated, with $p_{\mathrm{T}} > 15\,\mathrm{GeV}$ and $|\eta| < 2.4$, and to be identified as passing tight identification tracking criteria recommended by the CMS Muon Physics Object Group. Corrections are also applied to scale the raw energy measurements and smear the muon resolutions to match the accuracy and precision of reconstructed muons in MC simulated data to those in recorded data.

A.3.3 Jets

Particle candidates found by the PF algorithm are clustered into jets using the anti-k_T algorithm with distance parameter $R = 0.4$ (AK4), as discussed in Sect. 5.2.6. Selected jets should have $p_{\mathrm{T}} > 30\,\mathrm{GeV}$, $|\eta| < 2.4$, and pass tight jet identification requirements recommended by CMS Jet-MET Physics Object Group. In order to remove the overlapping of selected jets with the selected leptons, jet objects within $\Delta R_{\mathrm{jet,lepton}} < 0.4$ of selected leptons are not counted, where ΔR is defined in Sect. 2.2.1.

A.3.4 Missing Transverse Energy

In order to reduce the instrumental noise in the detector, $E_{\mathrm{T}}^{\mathrm{miss}}$ filters are applied as recommended by CMS Jet-MET Physics Object Group. An event selection requirement of $E_{\mathrm{T}}^{\mathrm{miss}} > 100$ is included because of the high threshold of $E_{\mathrm{T}}^{\mathrm{miss}}$ triggers.

A.3.5 Event Selection Criteria

Events with exactly two high-p_{T} isolated PF leptons (electrons or muons) with opposite electric charge are selected. Both electrons and muons are required to have $|\eta| < 2.4$, with $p_{\mathrm{T}} > 25\,\mathrm{GeV}$ ($p_{\mathrm{T}} > 15\,\mathrm{GeV}$) for the leading (sub-leading) lepton. The invariant mass of the two leptons is required to be greater than 20 GeV. Selected events are required to have at least one jet passing the loose working point of the DeepCSV b-tagging algorithm.

A.4 Statistical and Systematic Uncertainties

For the two-dimensional trigger efficiency scale factor measurements, the systematic uncertainties are added in quadrature with the statistical uncertainties for the final total uncertainties in each bin. For the one-dimensional trigger efficiency scale factor measurements, the sums in quadrature of the systematic uncertainties are plotted as a hashed uncertainty band for each bin, while the statistical uncertainties are shown as an independent error bar on the data points.

A.4.1 Statistical Uncertainties

The asymmetric statistical uncertainty for measured trigger efficiencies is determined using Clopper-Pearson intervals [1]. This is generally a conservative method because Clopper-Pearson intervals have the property that the actual coverage is no smaller than the nominal for any population portion. It is also described in the statistics section of the Particle Data Group (PDG) book [2].

The asymmetric statistical uncertainty for measured scale factors $T = (X/m)/(Y/n)$ is determined using the method recommended for ratios of efficiencies, in which $\ln(T)$ is approximately normally distributed [3]. The estimated variance is $\hat{\sigma}^2 = (1/x) - (1/m) + (1/y) - (1/n)$ and an approximate two-sided $1 - \alpha$ confidence interval (1σ : $0.6827 = 1 - 0.3173$) is given by $\left\{t \exp\left(-\xi_{1-\alpha/2} \cdot \hat{\sigma}\right), t \exp\left(\xi_{1-\alpha/2} \cdot \hat{\sigma}\right)\right\}$ where $\xi_{1-\alpha/2}$ is the $1 - \frac{1}{2}\alpha$ fractile of the standard normal distribution, and t, x, and y are the observed values of T, X, and Y respectively. For the extreme cases in which $x = m$ and $y = n$, x and y are replaced with $x = m - 1/2$ and $y = n - 1/2$.

A.4.2 Systematic Uncertainties

The sources of systematic uncertainties are related to event topology, dependence on era, and correlations between MET and dilepton triggers. In order to estimate the effect of event topology on the trigger scale factors, the data and MC samples are partitioned into two mutually exclusive regions using the number of jets and the number of vertices, and scale factors are computed separately in each region. An envelope of the largest bin-by-bin differences with respect to the nominal is taken from the two partitions for the number of jets and the number of vertices independently. The final event topology systematic uncertainty in each bin is the sum in quadrature of those two independent envelopes. The number of jets distributions is relatively stable for different conditions during Run II, and the number of jets partition is $[< 3, \geq 3]$ for all three eras. However, the number of primary vertices distributions are more strongly dependent on the luminosity conditions and a different number of vertex partition is chosen for each era such that roughly the same percentage of MC events are below the partition. For 2016, 2017, and 2018, the number of vertices partition are $[< 20, \geq 20]$, $[< 30, \geq 30]$, and $[< 30, \geq 30]$, respectively. The bin-by-bin differences between the nominal measurement of the scale factor and a luminosity-weighted combination of scale factors measured for the individual runs in each era is taken as another source of systematic uncertainty.

We estimate the correlation factor (α) between MET and dilepton (DL) triggers in $t\bar{t}$ MC events by finding the fraction of events passing only the MET triggers ($\varepsilon^{\text{MC}}_{\text{MET triggers}}$), only the dilepton triggers ($\varepsilon^{\text{MC}}_{\text{Dilepton triggers}}$) and passing both of them ($\varepsilon^{\text{MC}}_{\text{Dilepton triggers, MET triggers}}$):

$$\alpha = \frac{\varepsilon^{\text{MC}}_{\text{Dilepton triggers}} \times \varepsilon^{\text{MC}}_{\text{MET triggers}}}{\varepsilon^{\text{MC}}_{\text{Dilepton triggers \& MET triggers}}}. \tag{A.3}$$

If $E_{\text{T}}^{\text{miss}}$ and dilepton triggers are independent with no correlation, then $\varepsilon^{\text{MC}}_{\text{Dilepton triggers \& MET triggers}} = \varepsilon^{\text{MC}}_{\text{Dilepton triggers}} \times \varepsilon^{\text{MC}}_{\text{MET triggers}}$ or $\alpha = 1$ ($P(A \cap B) = P(A) \times P(B)$). The fraction ($\alpha$) is determined for each channel and data set from Eq. (A.3) and its difference from 1 is found to be small for all triggers studied. Because these values are negligible with respect to the other systematic uncertainties, they are not added in quadrature for the final total uncertainty. Values of α as a function of leading lepton p_{T} and sub-leading lepton p_{T} in the e^+e^- and

$\mu^+\mu^-$ channels and as a function of electron p_T and muon p_T in the $e^\pm\mu^\mp$ channel are shown in Figs. A.5, A.10, A.15, and A.20.

A.5 Results

A.5.1 Trigger Efficiencies and Scale Factors: 2016preVFP

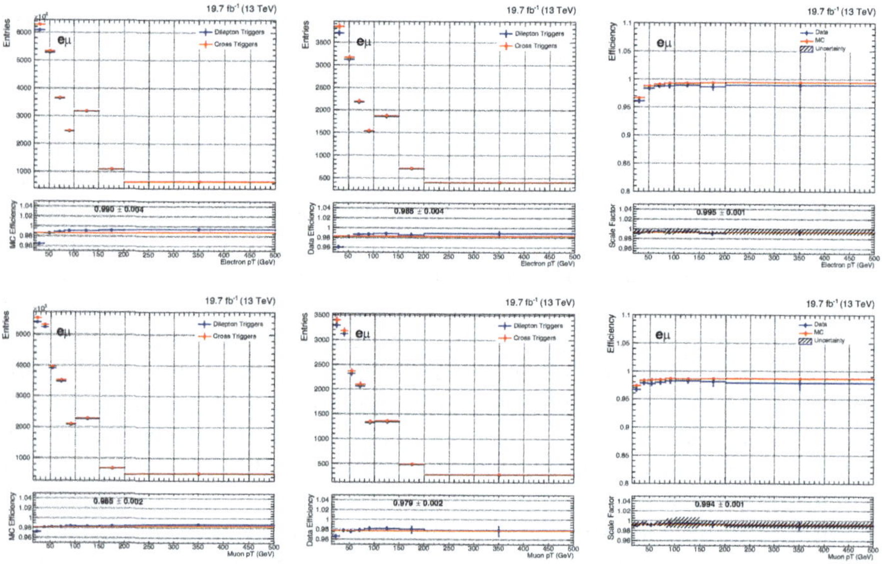

Fig. A.1 Efficiencies and scale factors for the 2016preVFP data set in the $e^\pm\mu^\mp$ channel as a function of electron and muon p_T. The error bars indicate the statistical uncertainty, and the hatched band corresponds to the systematic uncertainty

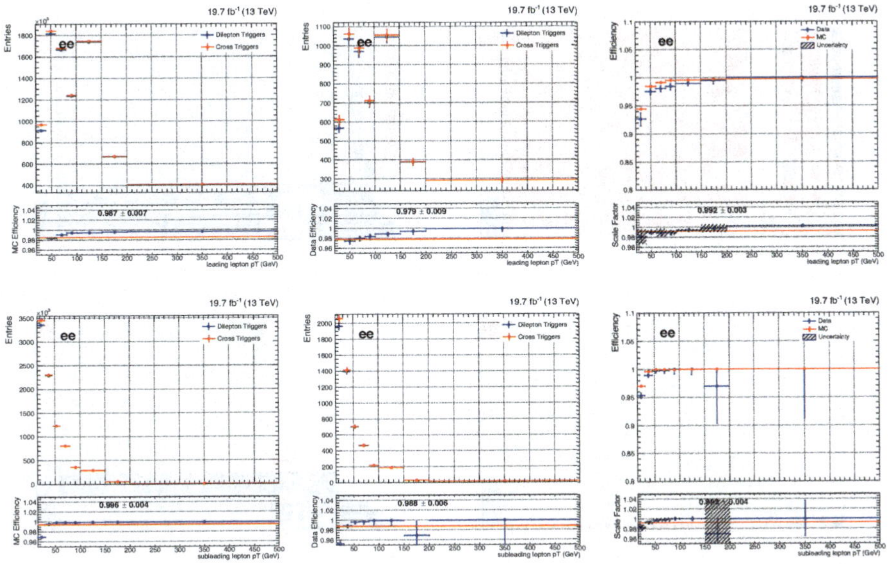

Fig. A.2 Efficiencies and scale factors for the 2016preVFP data set in the e^+e^- channel as a function of leading and sub-leading lepton p_T. The error bars indicate the statistical uncertainty, and the hatched band corresponds to the systematic uncertainty

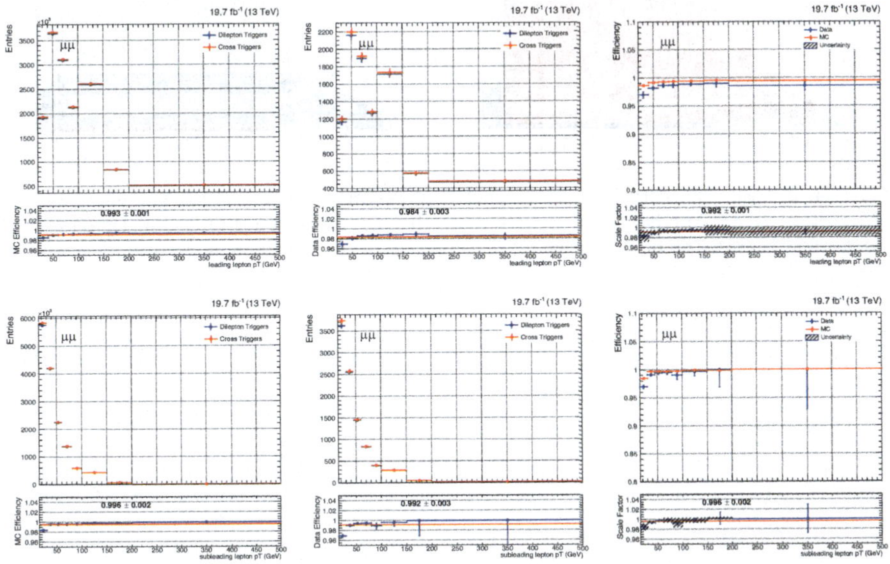

Fig. A.3 Efficiencies and scale factors for the 2016preVFP data set in the $\mu^+\mu^-$ channel as a function of leading and sub-leading lepton p_T. The error bars indicate the statistical uncertainty, and the hatched band corresponds to the systematic uncertainty

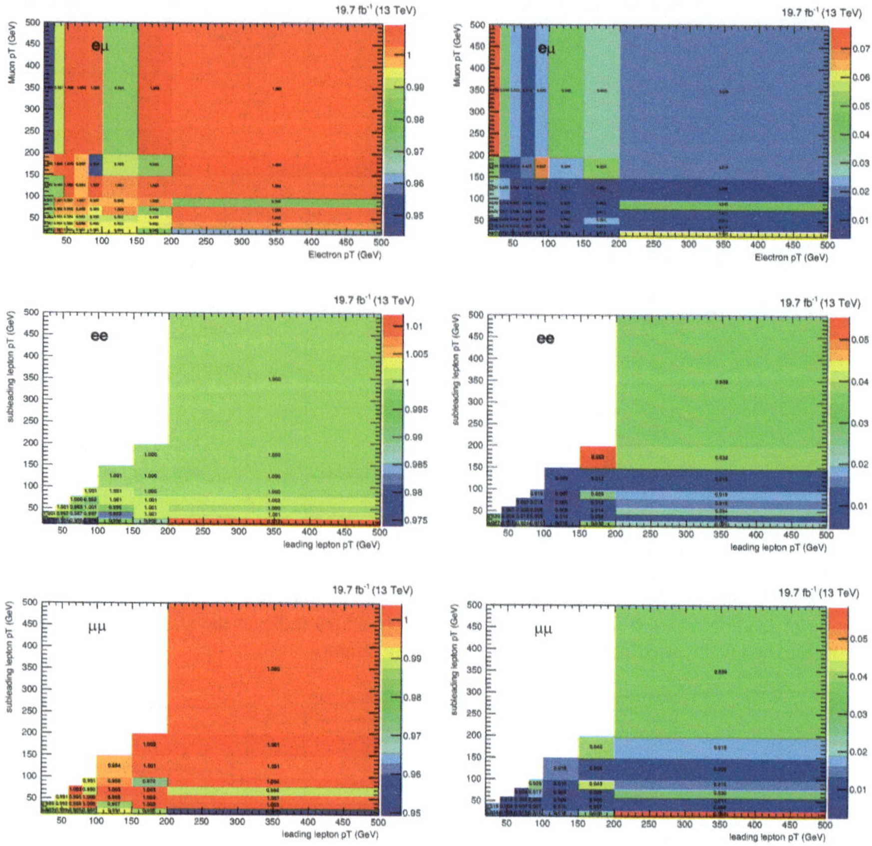

Fig. A.4 2D scale factors (Left) and total uncertainties (Right) for the 2016preVFP data set in the $e^{\pm}\mu^{\mp}$ (Top), $e^{+}e^{-}$ (Middle) and $\mu^{+}\mu^{-}$ (Bottom) channels as a function of leading lepton p_{T} and sub-leading lepton p_{T} in the $e^{+}e^{-}$ and $\mu^{+}\mu^{-}$ channels and as a function of electron p_{T} and muon p_{T} in the $e^{\pm}\mu^{\mp}$ channel

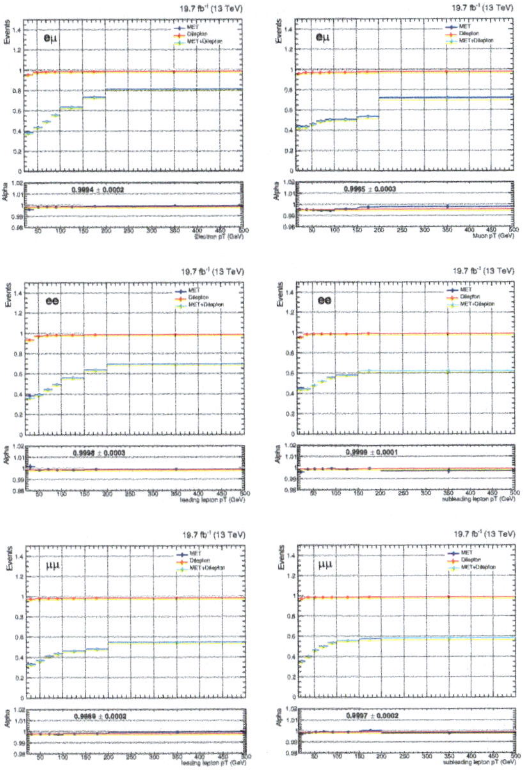

Fig. A.5 Correlations between MET and dilepton triggers for the 2016preVFP dataset in the $e^{\pm}\mu^{\mp}$ (Top), $e^{+}e^{-}$ (Middle) and $\mu^{+}\mu^{-}$ (Bottom) channels as a function of leading lepton p_{T} and sub-leading lepton p_{T} in the $e^{+}e^{-}$ and $\mu^{+}\mu^{-}$ channels and as a function of electron p_{T} and muon p_{T} in the $e^{\pm}\mu^{\mp}$ channel

A.5.2 Trigger Efficiencies and Scale Factors: 2016postVFP

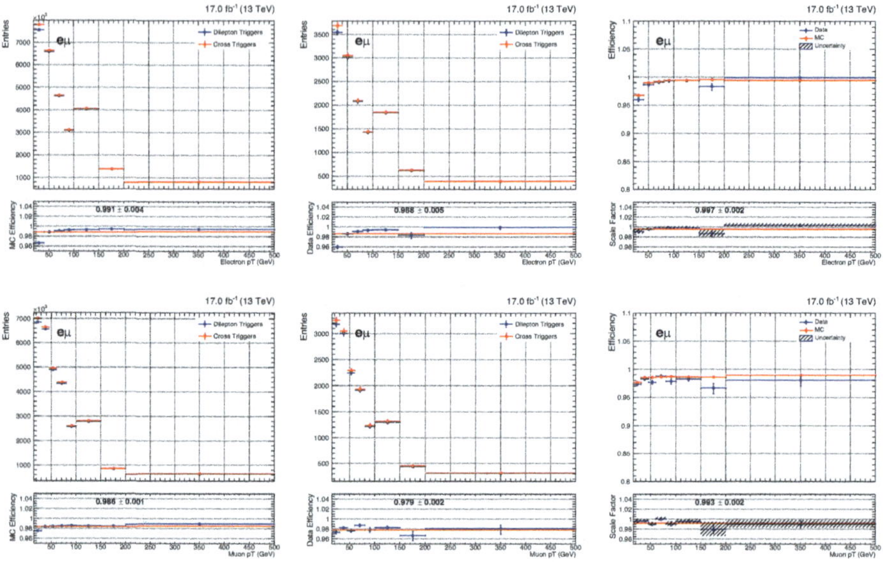

Fig. A.6 Efficiencies and scale factors for the 2016postVFP data set in the $e^{\pm}\mu^{\mp}$ channel as a function of electron and muon p_{T}. The error bars indicate the statistical uncertainty, and the hatched band corresponds to the systematic uncertainty

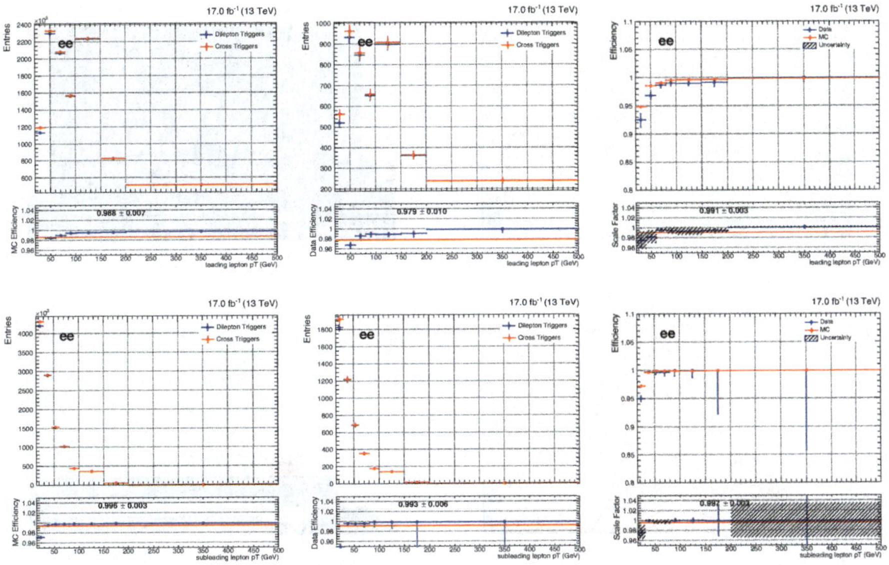

Fig. A.7 Efficiencies and scale factors for the 2016postVFP data set in the e^+e^- channel as a function of leading and sub-leading lepton p_T. The error bars indicate the statistical uncertainty, and the hatched band corresponds to the systematic uncertainty

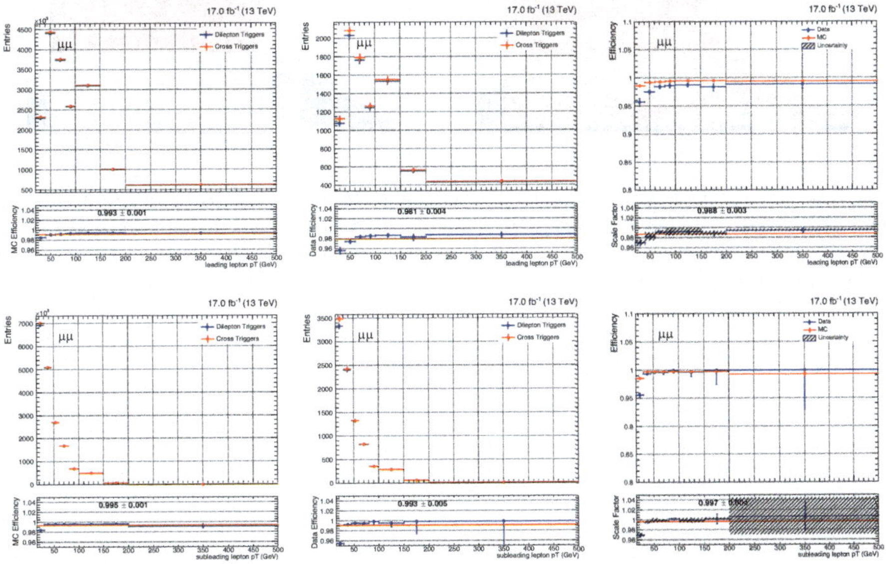

Fig. A.8 Efficiencies and scale factors for the 2016postVFP data set in the $\mu^+\mu^-$ channel as a function of leading and sub-leading lepton p_T. The error bars indicate the statistical uncertainty, and the hatched band corresponds to the systematic uncertainty

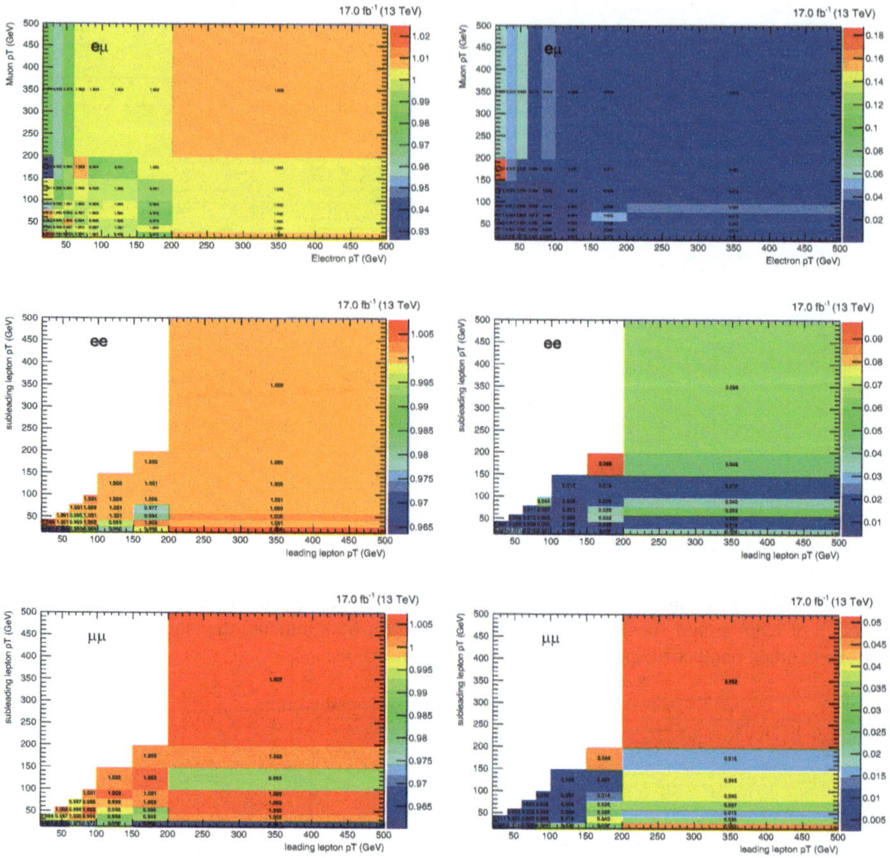

Fig. A.9 2D scale factors (Left) and total uncertainties (Right) for the 2016postVFP data set in the $e^{\pm}\mu^{\mp}$ (Top), $e^{+}e^{-}$ (Middle) and $\mu^{+}\mu^{-}$ (Bottom) channels as a function of leading lepton p_{T} and sub-leading lepton p_{T} in the $e^{+}e^{-}$ and $\mu^{+}\mu^{-}$ channels and as a function of electron p_{T} and muon p_{T} in the $e^{\pm}\mu^{\mp}$ channel

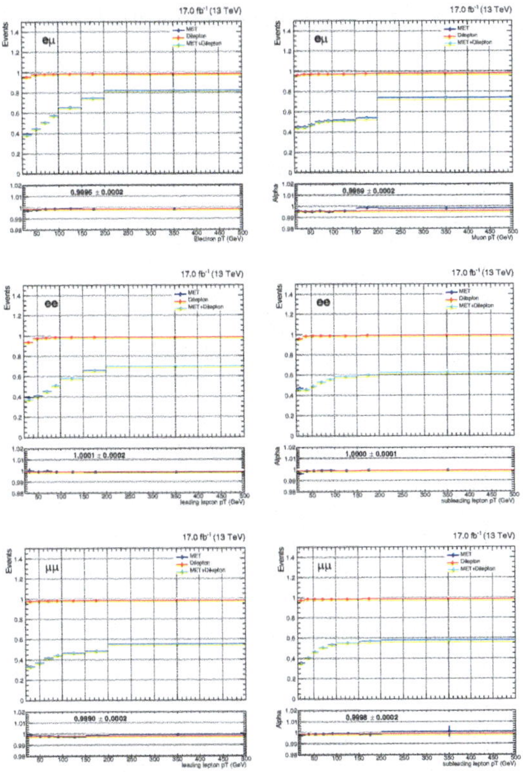

Fig. A.10 Correlations between MET and dilepton triggers for the 2016postVFP dataset in the $e^{\pm}\mu^{\mp}$ (Top), $e^{+}e^{-}$ (Middle) and $\mu^{+}\mu^{-}$ (Bottom) channels as a function of leading lepton p_{T} and sub-leading lepton p_{T} in the $e^{+}e^{-}$ and $\mu^{+}\mu^{-}$ channels and as a function of electron p_{T} and muon p_{T} in the $e^{\pm}\mu^{\mp}$ channel

A.5.3 Trigger Efficiencies and Scale Factors: 2017

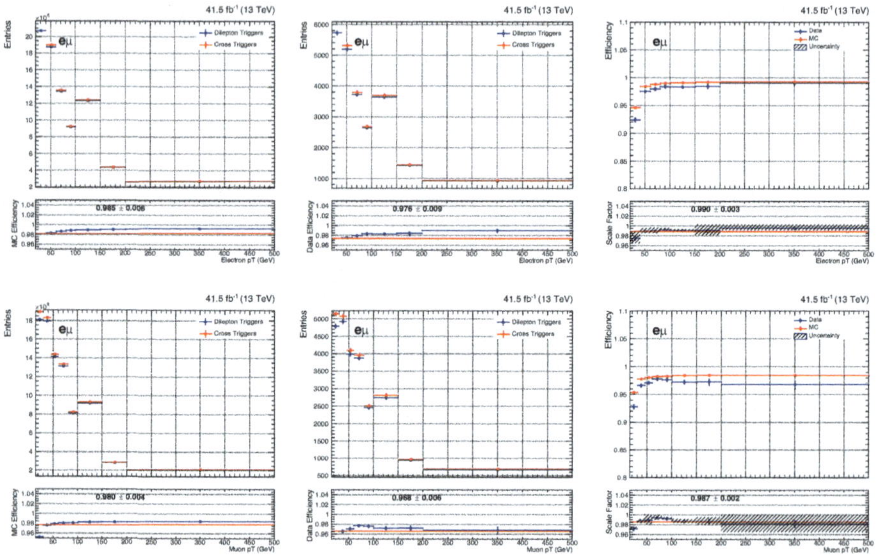

Fig. A.11 Efficiencies and scale factors for the 2017 data set in the $e^{\pm}\mu^{\mp}$ channel as a function of electron and muon p_{T}. The error bars indicate the statistical uncertainty, and the hatched band corresponds to the systematic uncertainty

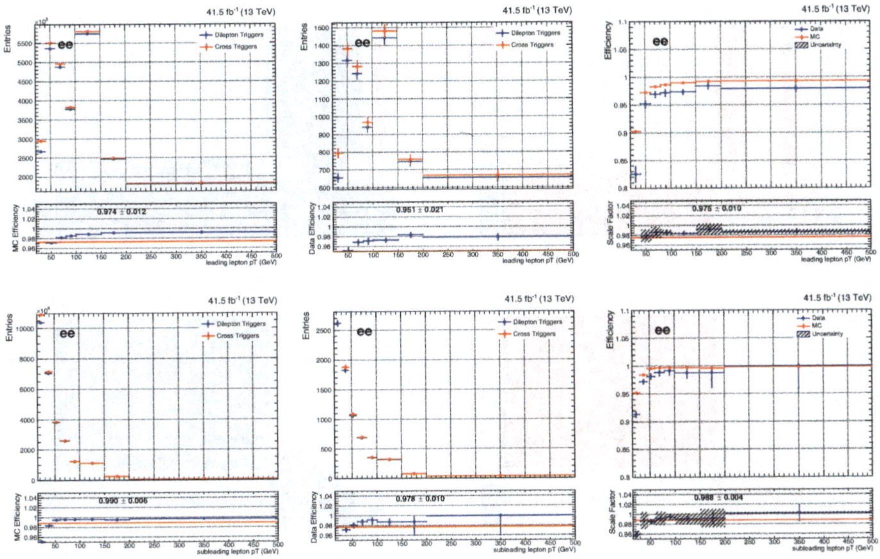

Fig. A.12 Efficiencies and scale factors for the 2017 data set in the e^+e^- channel as a function of leading and sub-leading lepton p_T. The error bars indicate the statistical uncertainty, and the hatched band corresponds to the systematic uncertainty

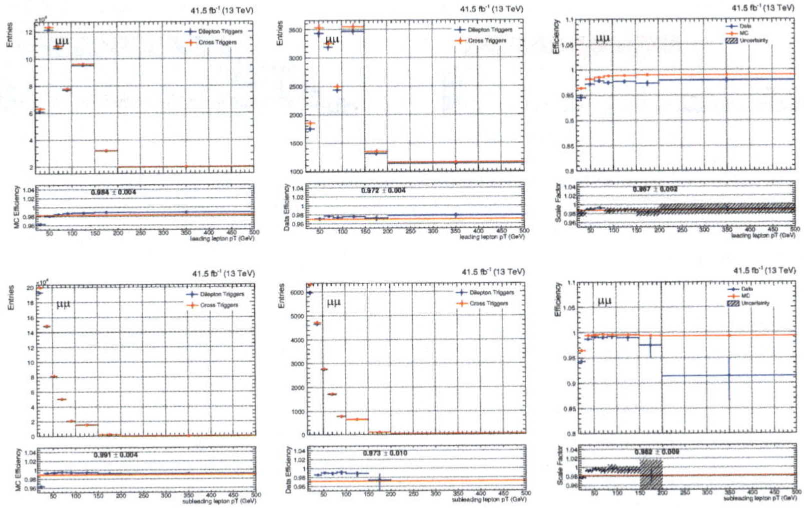

Fig. A.13 Efficiencies and scale factors for the 2017 data set in the $\mu^+\mu^-$ channel as a function of leading and sub-leading lepton p_T. The error bars indicate the statistical uncertainty, and the hatched band corresponds to the systematic uncertainty

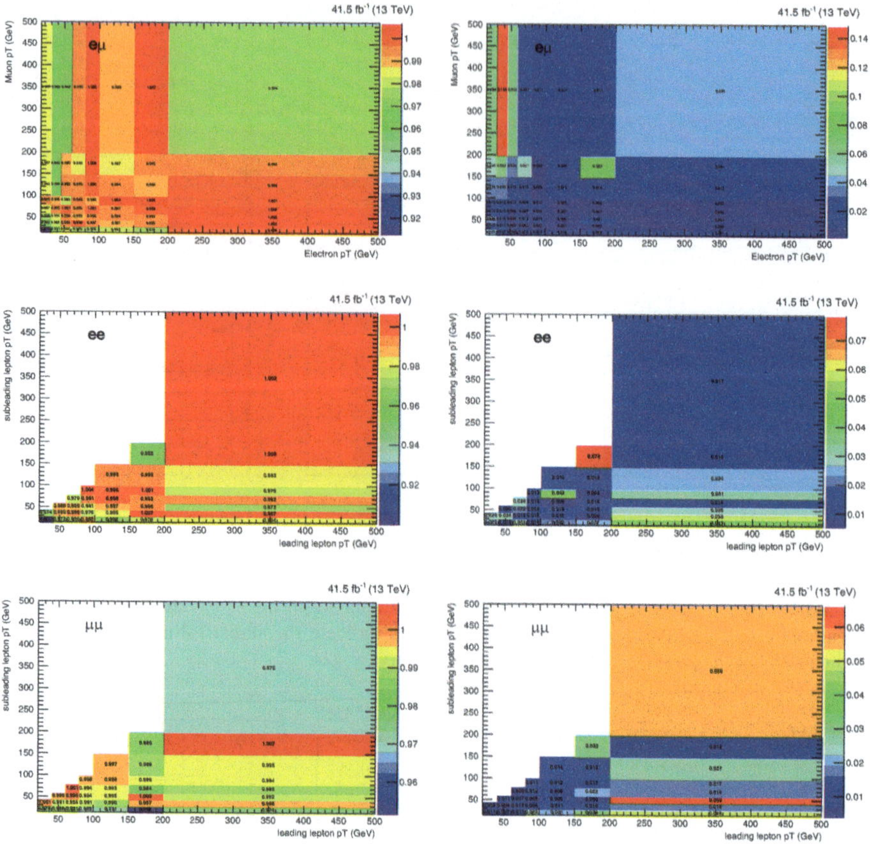

Fig. A.14 2D scale factors (Left) and total uncertainties (Right) for the 2017 data set in the $e^{\pm}\mu^{\mp}$ (Top), $e^{+}e^{-}$ (Middle) and $\mu^{+}\mu^{-}$ (Bottom) channels as a function of leading lepton p_{T} and subleading lepton p_{T} in the $e^{+}e^{-}$ and $\mu^{+}\mu^{-}$ channels and as a function of electron p_{T} and muon p_{T} in the $e^{\pm}\mu^{\mp}$ channel

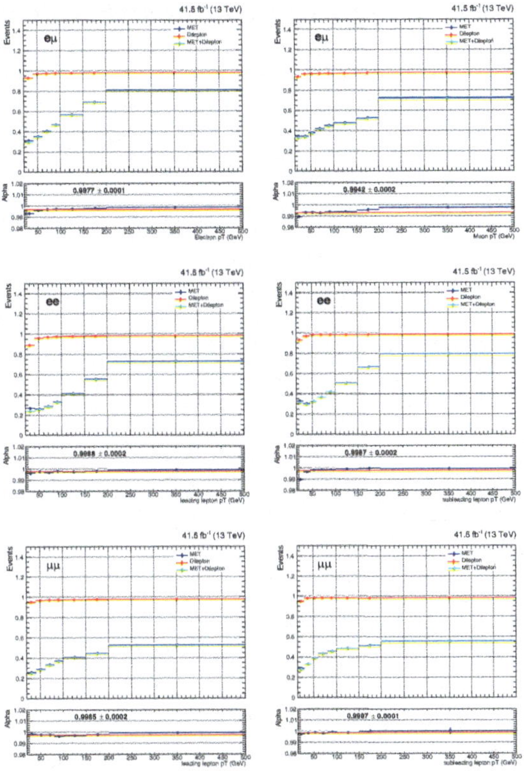

Fig. A.15 Correlations between MET and dilepton triggers for the 2017 dataset in the $e^{\pm}\mu^{\mp}$ (Top), $e^{+}e^{-}$ (Middle) and $\mu^{+}\mu^{-}$ (Bottom) channels as a function of leading lepton p_{T} and sub-leading lepton p_{T} in the $e^{+}e^{-}$ and $\mu^{+}\mu^{-}$ channels and as a function of electron p_{T} and muon p_{T} in the $e^{\pm}\mu^{\mp}$ channel

A.5.4 Trigger Efficiencies and Scale Factors: 2018

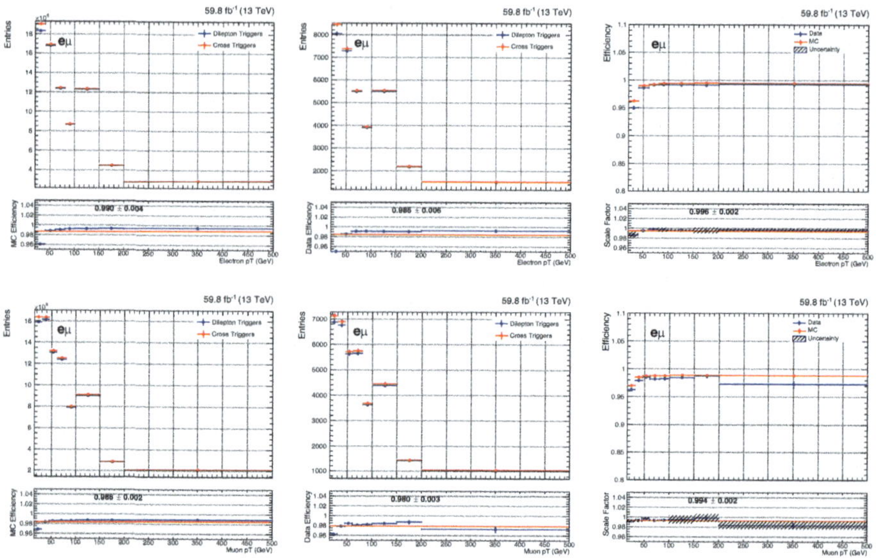

Fig. A.16 Efficiencies and scale factors for the 2018 data set in the $e^{\pm}\mu^{\mp}$ channel as a function of electron and muon p_{T}. The error bars indicate the statistical uncertainty, and the hatched band corresponds to the systematic uncertainty

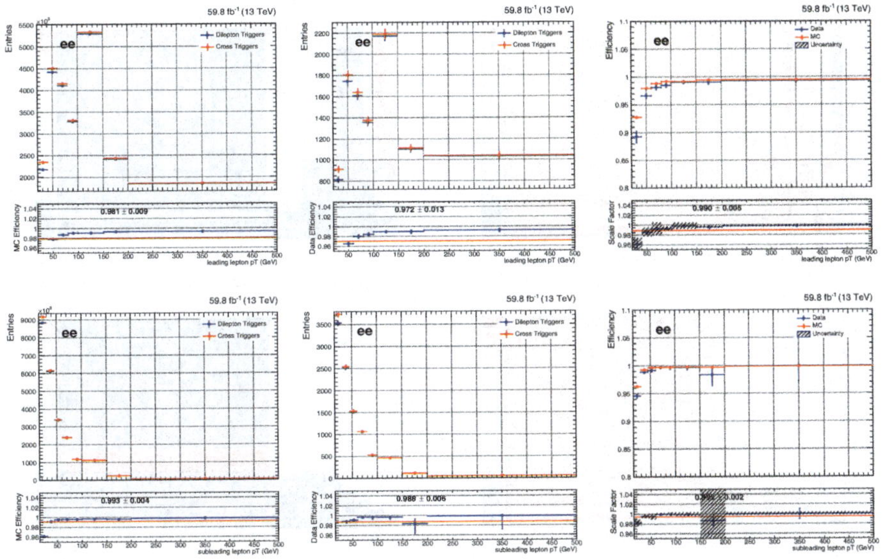

Fig. A.17 Efficiencies and scale factors for the 2018 data set in the e^+e^- channel as a function of leading and sub-leading lepton p_{T}. The error bars indicate the statistical uncertainty, and the hatched band corresponds to the systematic uncertainty

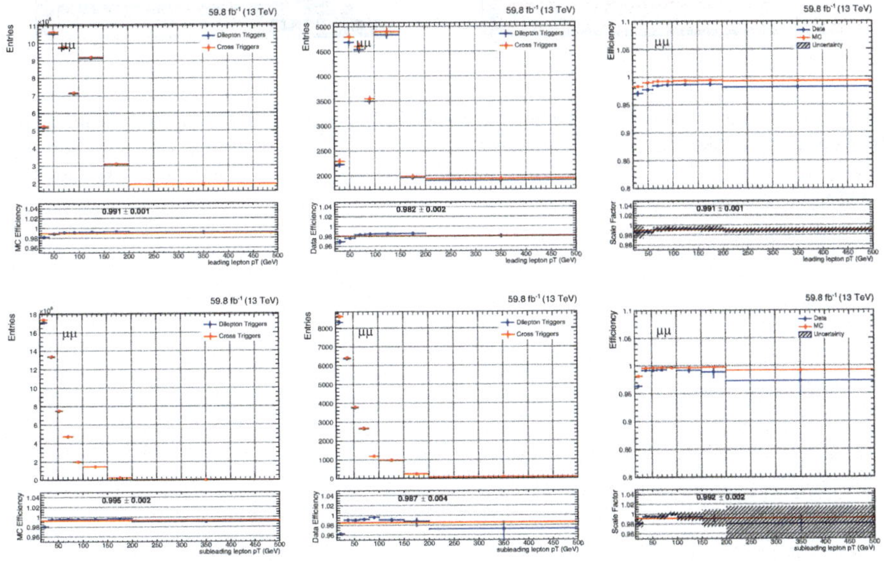

Fig. A.18 Efficiencies and scale factors for the 2018 data set in the $\mu^+\mu^-$ channel as a function of leading and sub-leading lepton p_{T}. The error bars indicate the statistical uncertainty, and the hatched band corresponds to the systematic uncertainty

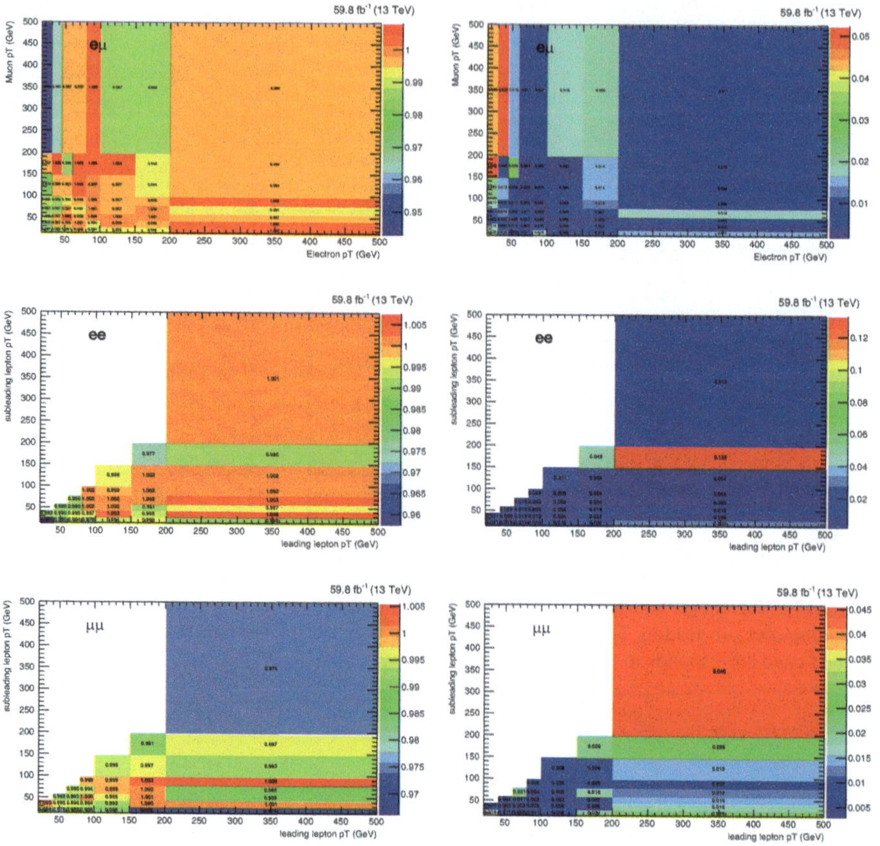

Fig. A.19 2D scale factors (Left) and total uncertainties (Right) for the 2018 data set in the $e^{\pm}\mu^{\mp}$ (Top), $e^{+}e^{-}$ (Middle) and $\mu^{+}\mu^{-}$ (Bottom) channels as a function of leading lepton p_{T} and subleading lepton p_{T} in the $e^{+}e^{-}$ and $\mu^{+}\mu^{-}$ channels and as a function of electron p_{T} and muon p_{T} in the $e^{\pm}\mu^{\mp}$ channel

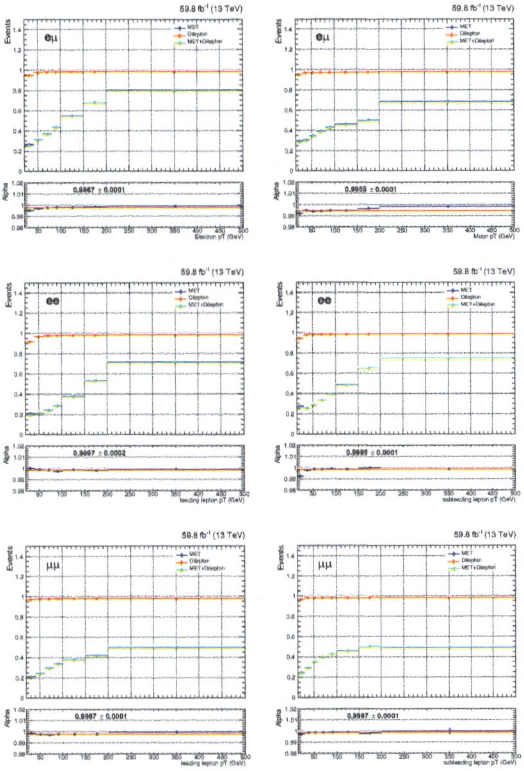

Fig. A.20 Correlations between MET and dilepton triggers for the 2018 dataset in the $e^{\pm}\mu^{\mp}$ (Top), $e^{+}e^{-}$ (Middle) and $\mu^{+}\mu^{-}$ (Bottom) channels as a function of leading lepton p_{T} and sub-leading lepton p_{T} in the $e^{+}e^{-}$ and $\mu^{+}\mu^{-}$ channels and as a function of electron p_{T} and muon p_{T} in the $e^{\pm}\mu^{\mp}$ channel

A.6 Search for Lorentz Invariance Violation in $t\bar{t}$ Production

Lorentz symmetry (i.e. that the laws of physics are the same in all inertial frames of reference) is a fundamental principle of general relativity and relativistic quantum field theories. Violation of Lorentz symmetry could indicate the presence of NP. One method to search for a violation of Lorentz invariance is to search for a modulation of the $t\bar{t}$ cross-section as a function of sidereal time (the orientation of the experiment relative to the fixed position of stars in the sky). The D0 collaboration performed such a search with semi-leptonic final states and the results were published in Physical Review Letters [4]. In order to accommodate a CMS search for a violation of Lorentz invariance (TOP-22-007), the dilepton trigger efficiency scale factors were measured one-dimensionally as a function of sidereal time using the same method outlined above, with the exception of using custom time-dependent corrections and not including the number of vertices partition as a source of event topology systematic uncertainty [5].

A.6.1 Translating UNIX Time to Sidereal Time

Every CMS data event is recorded with a UNIX timestamp. The sidereal day, which is the time it takes for one rotation of the Earth relative to the stars, is about 4 minutes shorter than the solar day, which is the time it takes for one rotation of the Earth relative to the Sun. In UNIX time, the rotational period of the earth (sidereal day) lasts approximately 23 h 56 min 4 s, i.e. 86,164 UNIX seconds. The sidereal second is defined in such a way that the rotational period of the earth is exactly 24 h, i.e. 86,400 sidereal seconds. The following formula is used to translate UNIX time to sidereal time:

$$\Omega_{sidereal} t_{sidereal} = \Omega_{UTC} \times (t_{UNIX} - t_0) + \phi_{UNIX} + \phi_{longitude} \tag{A.4}$$

In this equation:

- $\Omega_{sidereal}$ is the angular velocity of earth's rotation around its axis in sidereal time: $\Omega_{sidereal} = 2\pi/86,400\,\mathrm{s}^{-1}$ (sidereal).
- Ω_{UTC} is the angular velocity of earth's rotation around its axis in UTC time: $\Omega_{UTC} = 2\pi/86,164\,\mathrm{s}^{-1}$ (UTC).
- t_{UNIX} is the event timestamp recorded at CMS, in UNIX time; this is the number of UTC seconds since the UNIX epoch, i.e. the 1st of January 1970. However, handling large number of seconds is sometimes not practical and a t_0 origin is subtracted to set the origin at the first second of 2016.
- The azimuth angle ϕ_{UNIX} encodes the phase between the Unix epoch and "J2000"—the origin of sidereal time count. J2000 is defined as the direction pointed by the crossing of ecliptic and equator plans, at noon the 1st of January

2000, on the side of the earth were the Sun moves from the Southern hemisphere to the Northern hemisphere.

- The azimuth $\phi_{longitude}$ is the effective longitude of the beam at CMS relative to the Greenwich meridian, in radians.

A.6.2 Time Dependent Trigger Scale Factors: 2016

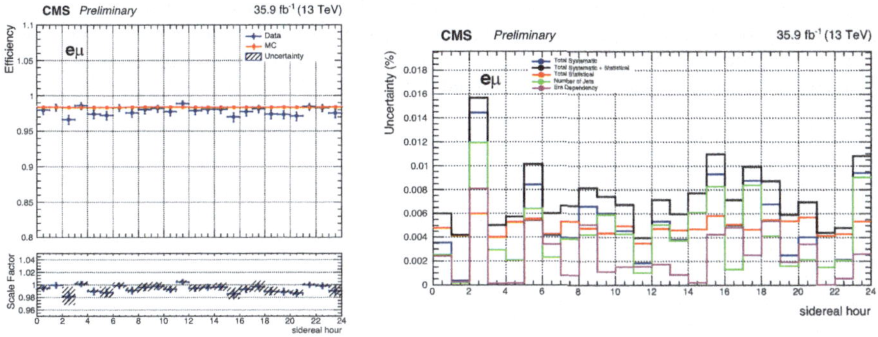

Fig. A.21 Trigger scale factors for the 2016 data set in the $e^{\pm}\mu^{\mp}$ channel as a function of sidereal time. The error bars indicate the statistical uncertainty, and the hatched band corresponds to the systematic uncertainty

A.6.3 Time Dependent Trigger Scale Factors: 2017

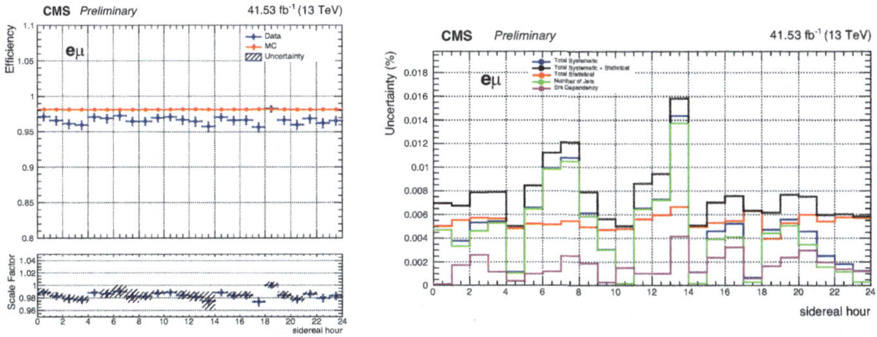

Fig. A.22 Trigger scale factors for the 2017 data set in the $e^{\pm}\mu^{\mp}$ channel as a function of sidereal time. The error bars indicate the statistical uncertainty, and the hatched band corresponds to the systematic uncertainty

References

1. R.D. Cousins, K.E. Hymes, J. Tucker, Frequentist evaluation of intervals estimated for a binomial parameter and for the ratio of Poisson means. Nucl. Instrum. Meth. **A612**, 388–398 (2010). https://doi.org/10.1016/j.nima.2009.10.156. arXiv: 0905.3831 [physics.data-an]
2. Particle Data Group, Review of particle physics. PTEP **2022**, 083C01 (2022). https://doi.org/10.1093/ptep/ptac097.
3. P.A.R. Koopman, Confidence intervals for the ratio of two binomial proportions. Bio-metrics **40**(2), 513–517 (1984). ISSN:0006341X, 15410420 [Online]. Available: http://www.jstor.org/stable/2531405
4. The D0 Collaboration, Search for violation of lorentz invariance in top quark pair production and decay. Phys. Rev. Lett. **108**, 261603 (2012). https://doi.org/10.1103/PhysRevLett.108.261603 [Online]. Available: https://link.aps.org/doi/10.1103/PhysRevLett.108.261603
5. CMS Collaboration, Searches for violation of Lorentz invariance in tt production using dilepton events in proton-proton collisions at $\sqrt{s} = 13$ TeV. CERN, Geneva, Technical Report, 2023 [Online]. Available: https://cds.cern.ch/record/2859658

Appendix B
Utilizing Induced Currents for Fast Timing Information in Pixel Detectors to Achieve Four-Dimensional Tracking

Appendix B explores the potential of using induced currents for fast timing information in pixel detectors, with the ultimate goal of achieving four-dimensional tracking. It begins with an overview and motivation for the research, explaining the need for precise timing and four-dimensional tracking in high-energy physics experiments. Results from simulations using TCAD and SPICE are presented, highlighting the promise of small pixels, induced currents, and fast electronics in terms of timing precision. The findings suggest that this novel concept has the potential to provide the required timing precision, and if realized, could lead to significant advancements in detector technology.

B.1 Overview and Motivation

To increase the physics reach potential of high-energy collision experiments, the HL-LHC upgrade aims to vastly increase the instantaneous luminosity of the LHC beyond its current value. As instantaneous collider luminosity increases, silicon pixel tracking detectors in HL-LHC experiments are faced with the challenge of resolving the trajectories of high-energy particles in environments with high pileup, rate, and radiation. With an expected pileup of up to 200 events per bunch crossing at the HL-LHC, with an average distance between vertices $\sim$ 500 µm and timing rms $\sim$ 150 ps [1], precise timing information would dramatically improve the reconstruction process by providing another dimension to separate vertices and avoid a loss of events due to pileup degradation.

To achieve precise timing measurements for HL-LHC collision, the CMS collaboration has designed the MIP timing sub-detector (MTD) to measure arrival times of high-energy particles [2]. The hybrid design of the MTD places LYSO scintillating crystals read-out by silicon photo-multipliers (SiPM) in the barrel, as well as low gain avalanche diodes (LGADs) in the end-caps, between the silicon

J. R. Thieman, *Top Quark Spin Correlations with the CMS Detector*,
Springer Theses, https://doi.org/10.1007/978-3-031-76117-1

tracker and ECAL layers, to complement timing information provided by the ECAL and provide time-of-flight information with timing resolutions on the order of a few tens of picoseconds.

Hybrid silicon pixel detectors consisting of silicon sensors bump-bonded to read-out chips currently constitute state-of-the-art silicon tracking detector technology and are the front-runners for track reconstruction and vertex identification in the high instantaneous luminosity environment of the HL-LHC. While these detectors have high granularity for spatial measurements, the timing resolution is relatively poor, because the sensors and readout electronics have cumbersome interconnects with high capacitance additions to the pre-amplifiers. Pixel detectors able to provide four-dimensional information for tracking have not yet been realized.

B.2 Small Pixels, Signal, and Noise

Collaborating with FNAL scientist, Dr. Ron Lipton, we explore whether small, low-capacitance pixels integrated with fast electronics show promise for realizing the timing resolution required for four-dimensional tracking. Smaller pixels provide better resolution, have better efficiency by avoiding in-pixel pileup, and have higher radiation tolerance due to less leakage current per pixel [3].

The timing resolution (σ_t) in a system limited by jitter (i.e. early or late firing fluctuations due to the presence of noise as illustrated in Fig. B.1) can be expressed as:

$$\sigma_t = \frac{\sigma_n}{\frac{dV}{dt}|_{V_T}} \approx t_r \frac{Noise}{Signal} \tag{B.1}$$

where σ_n is the front-end system noise, $\frac{dV}{dt}|_{V_T}$ is the slope of the signal at the threshold, and t_r is the rise time of the signal to the threshold crossing [5]. The system noise is proportional to the load capacitance, and the inverse square root

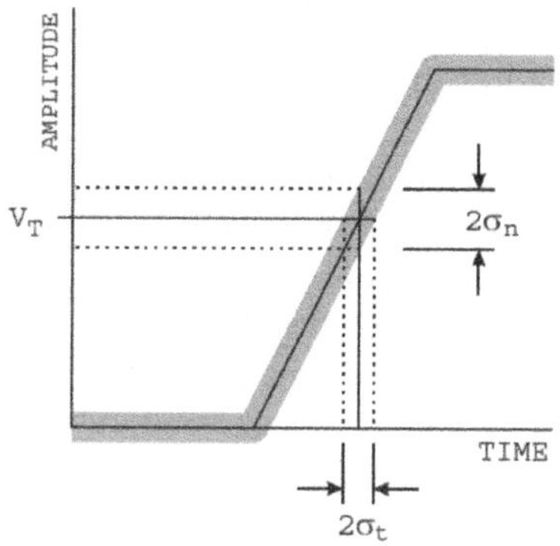

Fig. B.1 Fluctuations in signal amplitude crossing a voltage threshold translate into timing fluctuations called "jitter" [4]

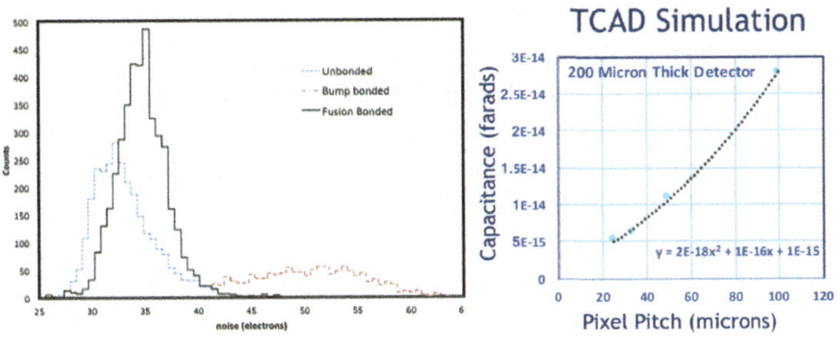

Fig. B.2 The noise in hybrid bonded VIPIC three-dimensional assembly is almost a factor of two lower than the equivalent conventionally bump bonded parts due to lower C_L (Left). TCAD simulations predict that a smaller pixel pitch results in lower detector capacitance (Right)

of the input transistor conductance, i.e. $\sigma_n \propto \frac{C_L}{g_m}$. Therefore, maximizing signal to noise, minimizing rise times t_r, increasing g_m, and minimizing load capacitance C_L yield better timing resolutions. The LGAD approach to this challenge is to use moderate gain to increase the signal [6], but we consider the approach of minimizing rise times and load capacitance.

B.3 Three-Dimensional Integration

For the noise component of minimizing jitter, three-dimensional integrated circuit (3DIC) technologies show promise for efforts towards realizing the timing resolution required for four-dimensional tracking by utlizing fine pitch, low capacitance bonding and interconnects to replace bump-bonding for hybrid silicon pixel detectors [7, 8]. The hybrid Ziptronix direct bonding interconnect (DBI) technique bonds planarized silicon wafers and electrically connects layers via the insertion of through silicon vias (TSV). Due to lower load capacitance, a hybrid, (DBI) bonded VIPIC three-dimensional prototype assembly demonstrated nearly a factor of two decrease in noise (see Fig. B.2) relative to a bump-bonded assembly [9]. Moreover, TCAD simulations predict that a smaller pixel pitch results in lower detector capacitance. In general, 3DIC technologies have many potential implementations for integrating silicon sensors with dense readout electronics, with advantages including radiation hardness, low mass, and decreasing costs, that could make them ideally suited for application in high-energy particle silicon pixel detectors.

B.4 Induced Currents for Fast Timing

Signal current, as registered by an electrode, begins before any charge is collected by the electrode. Current is instantaneously induced in the electrode as electrostatic flux lines ending on the electrode change due to charges moving in the vicinity [10]. According to the Ramo-Shockley theorem, current induced on an electrode in a multi-electrode system can be calculated from the coupling between the moving charge and the electrode, using a "weighting field," depending on geometry and doping, that is determined by setting the potential of the electrode under consideration to unity, and setting the potential of all other electrodes to zero [11]. The instantaneous current registered by an electrode by a charge moving in the vicinity can be expressed in terms of charge q, velocity $\vec{v}$, and weighting field $\vec{E_W}$ as:

$$i = -q\vec{v} \cdot \vec{E_W} \tag{B.2}$$

Induced currents have a very fast rise time, but if the charge is not collected by the electrode under consideration, then the induced signal current will change sign and integrate to zero. For geometries where the pixel pitch is small compared to the substrate thickness, a charge moving far from the surface will induce currents in a cluster of electrodes surrounding the nodes that collect charge. If detectors have sufficiently low capacitance and are integrated with fast electronics, the induced current signal can be utilized to provide timing information for four-dimensional tracking.

B.5 FEA Simulations with TCAD

Technology computer-aided design (TCAD) software is used to explore models for planar sensors in order to optimize the doping and geometry to achieve large induced currents with fast rise times. The TCAD simulation uses finite element analysis (FEA) to simulate the electrical properties and behavior of semiconductor models, in this case by calculating the electric field in the model, the weighting fields of the individual electrodes, and the current induced at each electrode due to the motion of charge carriers in the substrate. Both three-dimensional (3×3) and two-dimensional (1×9) models of pixel arrays, consisting of aluminum electrodes coupled to a silicon substrate, are simulated Fig. B.3. Different pixel pitches (25–100 μm), substrate thicknesses (50–300 μm), doping schemes (n-on-n, n-on-p), and bias voltages (100–400 V) were considered for these simulations. Electrons are the preferred charge carrier, as they have higher mobility than holes and thus higher velocity, thereby inducing larger currents for the same electric field strength. Even in the absence of an externally applied voltage, every pn-junction has a "built-in" electric field between the p and n sides from the thermal diffusion of mobile charge carriers. For the doping scheme, n-on-n has a maximal electric field near the bottom side, whereas n-on-p has a maximal electric field near the top Fig. B.4. Intermediate p-regions between electrodes, called "p-stops," help shape the electric field and prevent unwanted diffusion of charges away from collecting electrodes.

Two kinds of pulses were injected into the bulk of the substrate, either localized at a specific depth to simulate an absorbed x-ray, or uniformly through the depth at an

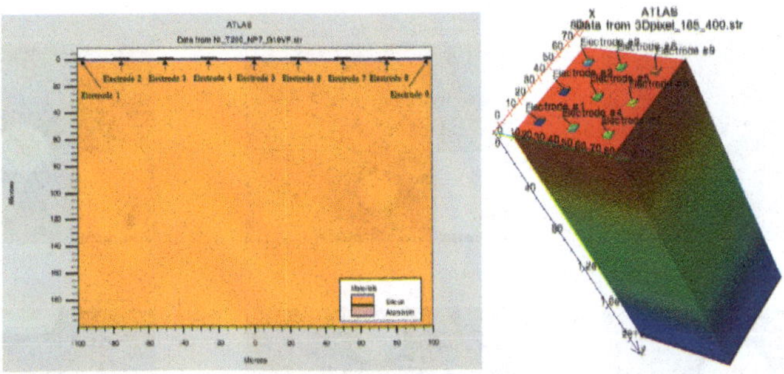

Fig. B.3 Two-dimensional 1×9 (Left) and three-dimensional 3×3 (Right) models of 25 μm pitch pixel arrays, consisting of aluminum electrodes (labeled) coupled to a silicon substrate

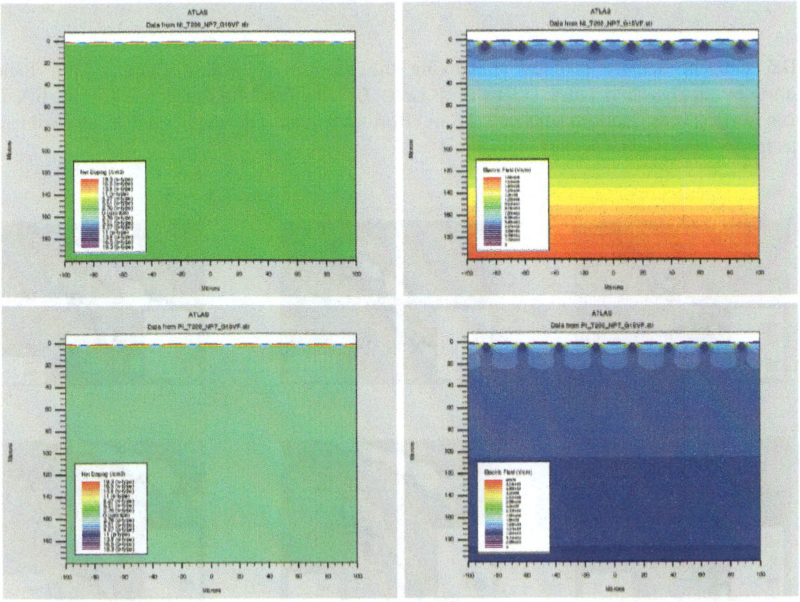

Fig. B.4 Doping (Left) and built-in electric field (Right) profiles for TCAD simulated models of 1×9 (two-dimensional) array of pixels with 25 μm pitch between aluminum electrodes coupled to a silicon substrate. n-on-n (Top) has an electric field maximal near the bottom side, whereas n-on-p (Bottom) has an electric field maximal near the top

angle of incidence to simulate ionization due to a MIP traversing the sensor. Current density profiles visualizing the motion of mobile charges after ionization are shown in Figs. B.5 and B.6 for an absorbed x-ray and MIP, respectively. Simulated signal currents induced on electrodes by a locally ionizing x-ray at various depths and by uniformly ionizing MIPs at various angles of incidence are shown in Figs. B.7 and B.8, respectively. The rich diversity of signal shapes in Fig. B.8 demonstrates that thick detectors with small pixels have potential for distinguishing tracks as a

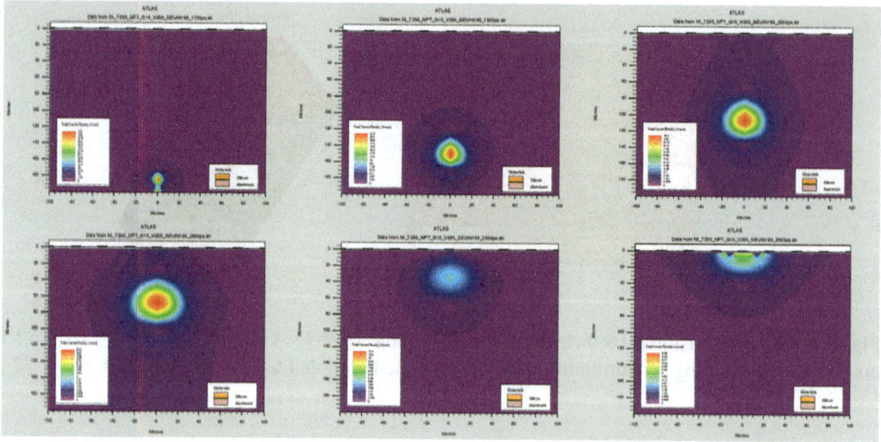

Fig. B.5 Current density profiles visualizing the motion of mobile charges after localized ionization by an absorbed x-ray: 0.1 ns (Top Left), 0.5 ns (Top Middle), 1 ns (Top Right), 1.5 ns (Bottom Left), 2.0 ns (Bottom Middle), 2.5 ns (Bottom Right). The electrons drift upward towards the electrodes, and the holes drift downward

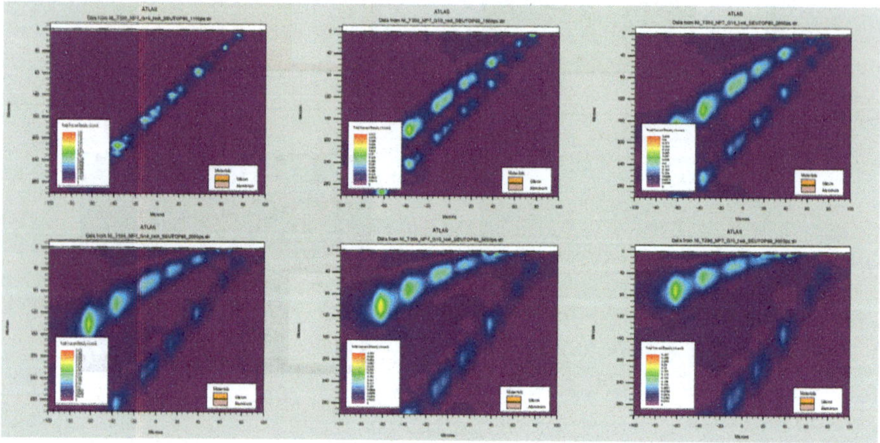

Fig. B.6 Current density profiles visualizing the motion of mobile charges after uniform ionization at an angle of incidence by a MIP: 0.1 ns (Top Left), 0.5 ns (Top Middle), 1 ns (Top Right), 1.5 ns (Bottom Left), 2.0 ns (Bottom Middle), 2.5 ns (Bottom Right). The electrons drift upward towards the electrodes, and the holes drift downward

function of the angle of incidence by using information from the collective pulse shapes of nearby pixels. The time between the initial current pulse and the peak current in a set of pixel shapes, for example, could be correlated in the readout electronics to measure the track angle.

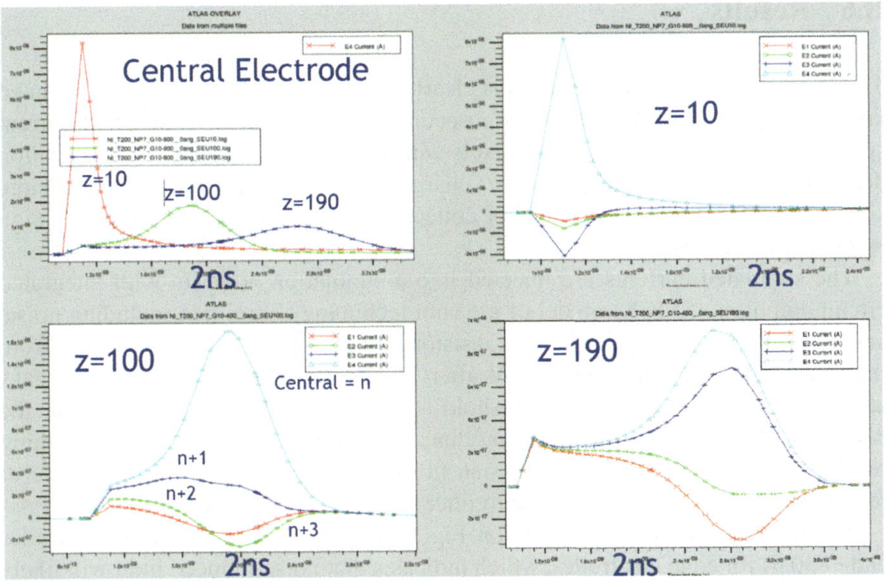

Fig. B.7 Currents induced on electrodes by a locally ionizing x-ray at various depths (10, 100, 190 μm) from the central electrode of a 200 μm thick detector with 25 μm pitch pixels biased at 400 V. The ionization is localized directly below Electrode 5 (E4)

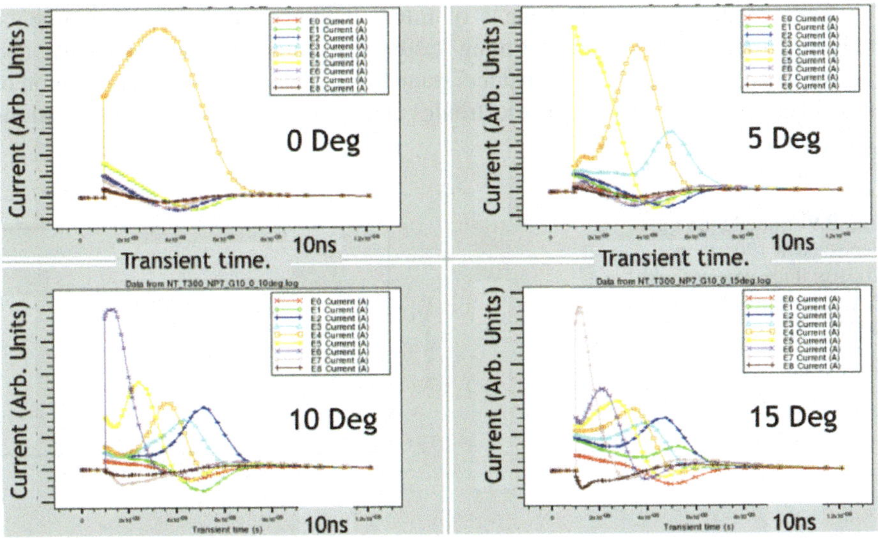

Fig. B.8 Currents induced on electrodes by a uniformly ionizing MIP at various angles of incidence with respect to the surface of a 200 μm thick detector with 25 μm pitch pixels biased at 400 V. The track at 0 degrees is incident on Electrode 5 (E4)

B.6 Results

The same pattern emerges that is most clearly demonstrated in the case of simulated x-ray absorption Fig. B.9. On the picosecond timescale, there is an initial current spike that is similar for all nearby pixels with a pitch small compared to ionization depth. On the nanosecond timescale, charges-collecting electrodes detect a large signal current, but for electrodes not collecting charges, the current changes sign and integrates to zero.

The simulated currents are injected into a simulation program with integrated circuit emphasis (SPICE) model of a 65 nm technology amplifier, including noise, and current supplied to the input transistor around 4 μA. The pulse shape from the central electrode peaks at $\sim$ 1 V after $\sim$ 12 ns, and the edge electrode peaks at $\sim$ 25 mV after $\sim$ 2 ns. The threshold is set at 130 mV above baseline for the central electrode and 10 mV above baseline for the edge electrode. The jitter timing resolution is extracted from a Gaussian fit of the arrival times of an ensemble of resulting pulses that differ only due to jitter noise.

The resulting jitter time resolution Fig. B.10 is $\sim$ 10 ps for central electrodes and $\sim$ 30 ps for edge electrodes, which indicates that noise-induced jitter will likely not be a dominant factor for timing resolution with small pixels. Time walk and ionization fluctuation components of timing resolution were not considered in this limited study.

As a concluding remark, small pixel detector systems show promise for addressing some of the extreme challenges of future high-energy particle experiments, including precise position and timing resolution, complex event topology, and radiation hardness. This study considers some toy models, but for real application of these ideas, there are many more electronics and system engineering considerations that need to be studied.

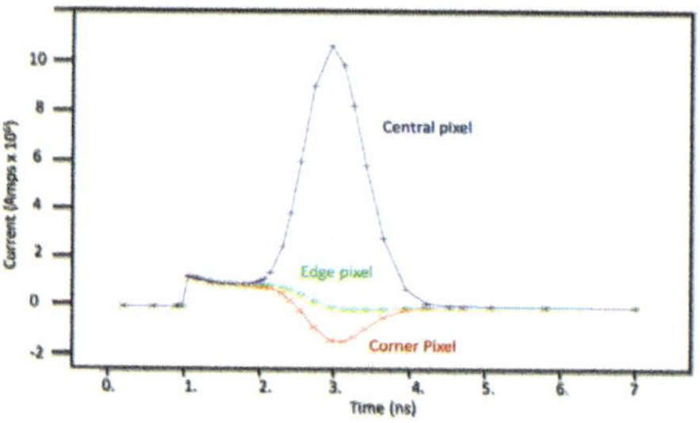

Fig. B.9 Signal for an X-ray deposited at 190 μm in a 200 μm thick detector with 25 μm pitch pixels biased at 400 V

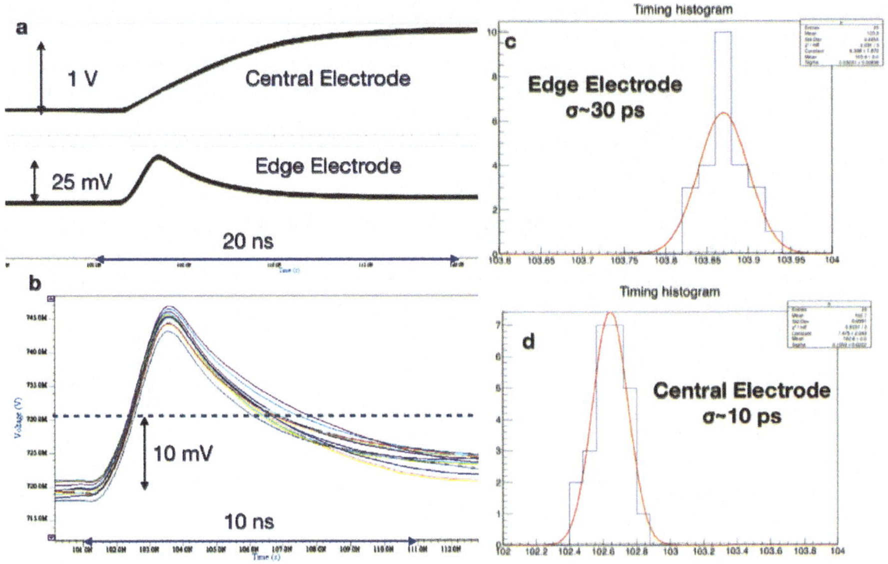

Fig. B.10 Pulse shapes and jitter timing resolution, of central (charge collecting) and edge (induced current) electrodes, for a simulated x-ray ionization at a depth of 185 μm in a 200 μm thick detector. (**a**) Pulse shapes for the central and edge electrode. (**b**) Pulse shape for the edge electrode for 25 pulses, including noise, with the threshold set at 10 mV above baseline. (**c**) The jitter timing resolution for the edge electrode in plot b. (**d**) The jitter timing resolution for the central electrode with a threshold of 130 mV above the baseline

References

1. N. Cartiglia, R. Arcidiacono, B. Baldassarri et al., Tracking in 4 dimensions. Nucl. Instru. Methods Phys. Res. Sect. A Accelerators, Spectrometers, Detectors Assoc. Equipment **845**, 47–51 (2017). Proceedings of the Vienna Conference on Instrumentation 2016. ISSN: 0168-9002. https://doi.org/10.1016/j.nima.2016.05.078 [Online]. Available: https://www.sciencedirect.com/science/article/pii/S0168900216304715
2. CMS Collaboration, A MIP Timing Detector for the CMS Phase-2 Upgrade. CERN, Geneva, Technical Report, 2019 [Online]. Available: https://cds.cern.ch/record/2667167
3. M. Garcia-Sciveres, N. Wermes, A review of advances in pixel detectors for experiments with high rate and radiation. Rep. Progress Phys. **81**(6), 066 101 (2018). https://doi.org/10.1088/1361-6633/aab064 [Online]. Available: https://dx.doi.org/10.1088/1361-6633/aab064
4. H. Spieler, *Semiconductor Detector Systems; 1st Publ.* (Series on Semiconductor Science and Technology), vol. 12 (Oxford University Press, Oxford, 2005), 489 p. ISBN: 0198527845 [Online]. Available: https://bib-pubdb1.desy.de/record/369858
5. H. Spieler, Fast timing methods for semiconductor detectors. IEEE Trans. Nucl. Sci. **29**(3), 1142–1158 (1982). https://doi.org/10.1109/TNS.1982.4336333
6. H.F.-W. Sadrozinski, S. Ely, V. Fadeyev et al., Ultra-fast silicon detectors. Nucl. Instr. Methods Phys. Res. Sect. A Accelerators, Spectrometers, Detectors Assoc. Equipment **730**, 226–231 (2013). Proceedings of the 9th International Conference on Radiation Effects on Semiconductor Materials Detectors and Devices. ISSN: 0168-9002. https://doi.org/10.1016/j.nima.2013.06.033 [Online]. Available: https://www.sciencedirect.com/science/article/pii/S0168900213008589

7. G.W. Deptuch, G. Carini, P. Enquist et al., Fully 3-d integrated pixel detectors for x-rays. IEEE Trans. Electron Dev. **63**(1), 205–214 (2016). https://doi.org/10.1109/TED.2015.2448671
8. G.W. Deptuch, G. Carini, T. Collier et al., Results of tests of three-dimensionally integrated chips bonded to sensors. IEEE Trans. Nucl. Sci. **62**(1), 349–358 (2015). https://doi.org/10.1109/TNS.2014.2378784
9. R. Lipton, 3D integration of sensors and electronics. PoS Vertex 2017, 025 (2018). https://doi.org/10.22323/1.309.0025
10. W. Shockley, Currents to conductors induced by a moving point charge. J. Appl. Phys. **9**(10), 635–636 (1938). https://doi.org/10.1063/1.1710367 [Online]. Available: https://doi.org/10.1063/1.1710367
11. S. Ramo, Currents induced by electron motion. Proc. IRE **27**(9), 584–585 (1939). https://doi.org/10.1109/JRPROC.1939.228757

SPRINGER NATURE

GPSR Compliance

The European Union's (EU) General Product Safety Regulation (GPSR) is a set of rules that requires consumer products to be safe and our obligations to ensure this.

If you have any concerns about our products, you can contact us on ProductSafety@springernature.com

In case Publisher is established outside the EU, the EU authorized representative is:

Springer Nature Customer Service Center GmbH
Europaplatz 3
69115 Heidelberg, Germany

The manufacturer's authorised representative in the EU is Springer Nature Customer Service Centre GmbH, Europaplatz 3, 69115 Heidelberg, Germany. If you have any concerns regarding our products, please contact ProductSafety@springernature.com

Printed and bound by CPI Group (UK) Ltd, Croydon, CR0 4YY

01/10/2026

02225916-0018